Pitman Research Notes in Mathematics Series

Submission of proposals for consideration

Suggestions for publication, in the form of outlines and representative samples, are invited by the Editorial Board for assessment. Intending authors should approach one of the main editors or another member of the Editorial Board, citing the relevant AMS subject classifications. Alternatively, outlines may be sent directly to the publisher's offices. Refereeing is by members of the board and other mathematical authorities in the topic concerned, throughout the world.

Preparation of accepted manuscripts

On acceptance of a proposal, the publisher will supply full instructions for the preparation of manuscripts in a form suitable for direct photo-lithographic reproduction. Specially printed grid sheets can be provided and a contribution is offered by the publisher towards the cost of typing. Word processor output, subject to the publisher's approval, is also acceptable.

Illustrations should be prepared by the authors, ready for direct reproduction without further improvement. The use of hand-drawn symbols should be avoided wherever possible, in order to maintain maximum clarity of the text.

The publisher will be pleased to give any guidance necessary during the preparation of a typescript, and will be happy to answer any queries.

Important note

In order to avoid later retyping, intending authors are strongly urged not to begin final preparation of a typescript before receiving the publisher's guidelines. In this way it is hoped to preserve the uniform appearance of the series.

Longman Scientific & Technical
Longman House
Burnt Mill
Harlow, Essex, CM20 2JE
UK
(Telephone (0279) 426721)

Titles in this series. A full list is available on request from the publisher.

51 Subnormal operators
J B Conway
52 Wave propagation in viscoelastic media
F Mainardi
53 Nonlinear partial differential equations and their applications: Collège de France Seminar. Volume I
H Brezis and J L Lions
54 Geometry of Coxeter groups
H Hiller
55 Cusps of Gauss mappings
T Banchoff, T Gaffney and C McCrory
56 An approach to algebraic K-theory
A J Berrick
57 Convex analysis and optimization
J-P Aubin and R B Vintner
58 Convex analysis with applications in the differentiation of convex functions
J R Giles
59 Weak and variational methods for moving boundary problems
C M Elliott and J R Ockendon
60 Nonlinear partial differential equations and their applications: Collège de France Seminar. Volume II
H Brezis and J L Lions
61 Singular Systems of differential equations II
S L Campbell
62 Rates of convergence in the central limit theorem
Peter Hall
63 Solution of differential equations by means of one-parameter groups
J M Hill
64 Hankel operators on Hilbert Space
S C Power
65 Schrödinger-type operators with continuous spectra
M S P Eastham and H Kalf
66 Recent applications of generalized inverses
S L Campbell
67 Riesz and Fredholm theory in Banach algebra
B A Barnes, G J Murphy, M R F Smyth and T T West
68 Evolution equations and their applications
K Kappel and W Schappacher
69 Generalized solutions of Hamilton–Jacobi equations
P L Lions
70 Nonlinear partial differential equations and their applications: Collège de France Seminar. Volume III
H Brezis and J L Lions
71 Spectral theory and wave operators for the Schrödinger equation
A M Berthier
72 Approximation of Hilbert space operators I
D A Herrero
73 Vector valued Nevanlinna theory
H J W Ziegler
74 Instability, nonexistence and weighted energy methods in fluid dynamics and related theories
B Straughan
75 Local bifurcation and symmetry
A Vanderbauwhede
76 Clifford analysis
F Brackx, R Delanghe and F Sommen
77 Nonlinear equivalence, reduction of PDEs to ODEs and fast convergent numerical methods
E E Rosinger
78 Free boundary problems, theory and applications. Volume I
A Fasano and M Primicerio
79 Free boundary problems, theory and applications. Volume II
A Fasano and M Primicerio
80 Symplectic geometry
A Crumeyrolle and J Grifone
81 An algorithmic analysis of a communication model with retransmission of flawed messages
D M Lucantoni
82 Geometric games and their applications
W H Ruckle
83 Additive groups of rings
S Feigelstock
84 Nonlinear partial differential equations and their applications: Collège de France Seminar. Volume IV
H Brezis and J L Lions
85 Multiplicative functionals on topological algebras
T Husain
86 Hamilton–Jacobi equations in Hilbert spaces
V Barbu and G Da Prato
87 Harmonic maps with symmetry, harmonic morphisms and deformations of metric
P Baird
88 Similarity solutions of nonlinear partial differential equations
L Dresner
89 Contributions to nonlinear partial differential equations
C Bardos, A Damlamian, J I Díaz and J Hernández
90 Banach and Hilbert spaces of vector-valued functions
J Burbea and P Masani
91 Control and observation of neutral systems
D Salamon
92 Banach bundles, Banach modules and automorphisms of C^*-algebras
M J Dupré and R M Gillette
93 Nonlinear partial differential equations and their applications: Collège de France Seminar. Volume V
H Brezis and J L Lions
94 Computer algebra in applied mathematics: an introduction to MACSYMA
R H Rand
95 Advances in nonlinear waves. Volume I
L Debnath
96 FC-groups
M J Tomkinson
97 Topics in relaxation and ellipsoidal methods
M Akgül
98 Analogue of the group algebra for topological semigroups
H Dzinotyiweyi
99 Stochastic functional differential equations
S E A Mohammed

100 Optimal control of variational inequalities
V Barbu
101 Partial differential equations and dynamical systems
W E Fitzgibbon III
102 Approximation of Hilbert space operators Volume II
C Apostol, L A Fialkow, D A Herrero and D Voiculescu
103 Nondiscrete induction and iterative processes
V Ptak and F-A Potra
104 Analytic functions – growth aspects
O P Juneja and G P Kapoor
105 Theory of Tikhonov regularization for Fredholm equations of the first kind
C W Groetsch
106 Nonlinear partial differential equations and free boundaries. Volume I
J I Díaz
107 Tight and taut immersions of manifolds
T E Cecil and P J Ryan
108 A layering method for viscous, incompressible L_p flows occupying R^n
A Douglis and E B Fabes
109 Nonlinear partial differential equations and their applications: Collège de France Seminar. Volume VI
H Brezis and J L Lions
110 Finite generalized quadrangles
S E Payne and J A Thas
111 Advances in nonlinear waves. Volume II
L Debnath
112 Topics in several complex variables
E Ramírez de Arellano and D Sundararaman
113 Differential equations, flow invariance and applications
N H Pavel
114 Geometrical combinatorics
F C Holroyd and R J Wilson
115 Generators of strongly continuous semigroups
J A van Casteren
116 Growth of algebras and Gelfand–Kirillov dimension
G R Krause and T H Lenagan
117 Theory of bases and cones
P K Kamthan and M Gupta
118 Linear groups and permutations
A R Camina and E A Whelan
119 General Wiener–Hopf factorization methods
F-O Speck
120 Free boundary problems: applications and theory. Volume III
A Bossavit, A Damlamian and M Fremond
121 Free boundary problems: applications and theory. Volume IV
A Bossavit, A Damlamian and M Fremond
122 Nonlinear partial differential equations and their applications: Collège de France Seminar. Volume VII
H Brezis and J L Lions
123 Geometric methods in operator algebras
H Araki and E G Effros
124 Infinite dimensional analysis–stochastic processes
S Albeverio
125 Ennio de Giorgi Colloquium
P Krée
126 Almost-periodic functions in abstract spaces
S Zaidman
127 Nonlinear variational problems
A Marino, L Modica, S Spagnolo and M Degliovanni
128 Second-order systems of partial differential equations in the plane
L K Hua, W Lin and C-Q Wu
129 Asymptotics of high-order ordinary differential equations
R B Paris and A D Wood
130 Stochastic differential equations
R Wu
131 Differential geometry
L A Cordero
132 Nonlinear differential equations
J K Hale and P Martinez-Amores
133 Approximation theory and applications
S P Singh
134 Near-rings and their links with groups
J D P Meldrum
135 Estimating eigenvalues with *a posteriori/a priori* inequalities
J R Kuttler and V G Sigillito
136 Regular semigroups as extensions
F J Pastijn and M Petrich
137 Representations of rank one Lie groups
D H Collingwood
138 Fractional calculus
G F Roach and A C McBride
139 Hamilton's principle in continuum mechanics
A Bedford
140 Numerical analysis
D F Griffiths and G A Watson
141 Semigroups, theory and applications. Volume I
H Brezis, M G Crandall and F Kappel
142 Distribution theorems of L-functions
D Joyner
143 Recent developments in structured continua
D De Kee and P Kaloni
144 Functional analysis and two-point differential operators
J Locker
145 Numerical methods for partial differential equations
S I Hariharan and T H Moulden
146 Completely bounded maps and dilations
V I Paulsen
147 Harmonic analysis on the Heisenberg nilpotent Lie group
W Schempp
148 Contributions to modern calculus of variations
L Cesari
149 Nonlinear parabolic equations: qualitative properties of solutions
L Boccardo and A Tesei
150 From local times to global geometry, control and physics
K D Elworthy

151 A stochastic maximum principle for optimal control of diffusions
U G Haussmann
152 Semigroups, theory and applications. Volume II
H Brezis, M G Crandall and F Kappel
153 A general theory of integration in function spaces
P Muldowney
154 Oakland Conference on partial differential equations and applied mathematics
L R Bragg and J W Dettman
155 Contributions to nonlinear partial differential equations. Volume II
J I Díaz and P L Lions
156 Semigroups of linear operators: an introduction
A C McBride
157 Ordinary and partial differential equations
B D Sleeman and R J Jarvis
158 Hyperbolic equations
F Colombini and M K V Murthy
159 Linear topologies on a ring: an overview
J S Golan
160 Dynamical systems and bifurcation theory
M I Camacho, M J Pacifico and F Takens
161 Branched coverings and algebraic functions
M Namba
162 Perturbation bounds for matrix eigenvalues
R Bhatia
163 Defect minimization in operator equations: theory and applications
R Reemtsen
164 Multidimensional Brownian excursions and potential theory
K Burdzy
165 Viscosity solutions and optimal control
R J Elliott
166 Nonlinear partial differential equations and their applications: Collège de France Seminar. Volume VIII
H Brezis and J L Lions
167 Theory and applications of inverse problems
H Haario
168 Energy stability and convection
G P Galdi and B Straughan
169 Additive groups of rings. Volume II
S Feigelstock
170 Numerical analysis 1987
D F Griffiths and G A Watson
171 Surveys of some recent results in operator theory. Volume I
J B Conway and B B Morrel
172 Amenable Banach algebras
J-P Pier
173 Pseudo-orbits of contact forms
A Bahri
174 Poisson algebras and Poisson manifolds
K H Bhaskara and K Viswanath
175 Maximum principles and eigenvalue problems in partial differential equations
P W Schaefer
176 Mathematical analysis of nonlinear, dynamic processes
K U Grusa
177 Cordes' two-parameter spectral representation theory
D F McGhee and R H Picard
178 Equivariant K-theory for proper actions
N C Phillips
179 Elliptic operators, topology and asymptotic methods
J Roe
180 Nonlinear evolution equations
J K Engelbrecht, V E Fridman and E N Pelinovski
181 Nonlinear partial differential equations and their applications: Collège de France Seminar. Volume IX
H Brezis and J L Lions
182 Critical points at infinity in some variational problems
A Bahri
183 Recent developments in hyperbolic equations
L Cattabriga, F Colombini, M K V Murthy and S Spagnolo
184 Optimization and identification of systems governed by evolution equations on Banach space
N U Ahmed
185 Free boundary problems: theory and applications. Volume I
K H Hoffmann and J Sprekels
186 Free boundary problems: theory and applications. Volume II
K H Hoffmann and J Sprekels
187 An introduction to intersection homology theory
F Kirwan
188 Derivatives, nuclei and dimensions on the frame of torsion theories
J S Golan and H Simmons
189 Theory of reproducing kernels and its applications
S Saitoh
190 Volterra integrodifferential equations in Banach spaces and applications
G Da Prato and M Iannelli
191 Nest algebras
K R Davidson
192 Surveys of some recent results in operator theory. Volume II
J B Conway and B B Morrel
193 Nonlinear variational problems. Volume II
A Marino and M K V Murthy
194 Stochastic processes with multidimensional parameter
M E Dozzi
195 Prestressed bodies
D Iesan
196 Hilbert space approach to some classical transforms
R H Picard
197 Stochastic calculus in application
J R Norris
198 Radical theory
B J Gardner
199 The C^*-algebras of a class of solvable Lie groups
X Wang
200 Stochastic analysis, path integration and dynamics
K D Elworthy and J C Zambrini

201 Riemannian geometry and holonomy groups
S Salamon
202 Strong asymptotics for extremal errors and polynomials associated with Erdös type weights
D S Lubinsky
203 Optimal control of diffusion processes
V S Borkar
204 Rings, modules and radicals
B J Gardner
205 Two-parameter eigenvalue problems in ordinary differential equations
M Faierman
206 Distributions and analytic functions
R D Carmichael and D Mitrovic
207 Semicontinuity, relaxation and integral representation in the calculus of variations
G Buttazzo
208 Recent advances in nonlinear elliptic and parabolic problems
P Bénilan, M Chipot, L Evans and M Pierre
209 Model completions, ring representations and the topology of the Pierce sheaf
A Carson
210 Retarded dynamical systems
G Stepan
211 Function spaces, differential operators and nonlinear analysis
L Paivarinta
212 Analytic function theory of one complex variable
C C Yang, Y Komatu and K Niino
213 Elements of stability of visco-elastic fluids
J Dunwoody
214 Jordan decomposition of generalized vector measures
K D Schmidt
215 A mathematical analysis of bending of plates with transverse shear deformation
C Constanda
216 Ordinary and partial differential equations. Volume II
B D Sleeman and R J Jarvis
217 Hilbert modules over function algebras
R G Douglas and V I Paulsen
218 Graph colourings
R Wilson and R Nelson
219 Hardy-type inequalities
A Kufner and B Opic
220 Nonlinear partial differential equations and their applications: Collège de France Seminar. Volume X
H Brezis and J L Lions
221 Workshop on dynamical systems
E Shiels and Z Coelho
222 Geometry and analysis in nonlinear dynamics
H W Broer and F Takens
223 Fluid dynamical aspects of combustion theory
M Onofri and A Tesei
224 Approximation of Hilbert space operators. Volume I. 2nd edition
D Herrero
225 Operator theory: proceedings of the 1988 GPOTS–Wabash conference
J B Conway and B B Morrel
226 Local cohomology and localization
J L Bueso Montero, B Torrecillas Jover and A Verschoren
227 Nonlinear waves and dissipative effects
D Fusco and A Jeffrey
228 Numerical analysis 1989
D F Griffiths and G A Watson
229 Recent developments in structured continua. Volume III
D De Kee and P Kaloni
230 Boolean methods in interpolation and approximation
F J Delvos and W Schempp
231 Further advances in twistor theory. Volume I
L J Mason and L P Hughston
232 Further advances in twistor theory. Volume II
L J Mason and L P Hughston
233 Geometry in the neighborhood of invariant manifolds of maps and flows and linearization
U Kirchgraber and K Palmer
234 Quantales and their applications
K I Rosenthal
235 Integral equations and inverse problems
V Petkov and R Lazarov
236 Pseudo-differential operators
S R Simanca
237 A functional analytic approach to statistical experiments
I M Bomze
238 Quantum mechanics, algebras and distributions
D Dubin and M Hennings
239 Hamilton flows and evolution semigroups
J Gzyl
240 Topics in controlled Markov chains
V S Borkar
241 Invariant manifold theory for hydrodynamic transition
S Sritharan
242 Lectures on the spectrum of $L^2(\Gamma\backslash G)$
F L Williams
243 Progress in variational methods in Hamiltonian systems and elliptic equations
M Girardi, M Matzeu and F Pacella
244 Optimization and nonlinear analysis
A Ioffe, M Marcus and S Reich
245 Inverse problems and imaging
G F Roach
246 Semigroup theory with applications to systems and control
N U Ahmed
247 Periodic-parabolic boundary value problems and positivity
P Hess
248 Distributions and pseudo-differential operators
S Zaidman
249 Progress in partial differential equations: the Metz surveys
M Chipot and J Saint Jean Paulin
250 Differential equations and control theory
V Barbu

251 Stability of stochastic differential equations with respect to semimartingales
X Mao

252 Fixed point theory and applications
J Baillon and M Théra

253 Nonlinear hyperbolic equations and field theory
M K V Murthy and S Spagnolo

254 Ordinary and partial differential equations. Volume III
B D Sleeman and R J Jarvis

255 Harmonic maps into homogeneous spaces
M Black

256 Boundary value and initial value problems in complex analysis: studies in complex analysis and its applications to PDEs 1
R Kühnau and W Tutschke

257 Geometric function theory and applications of complex analysis in mechanics: studies in complex analysis and its applications to PDEs 2
R Kühnau and W Tutschke

258 The development of statistics: recent contributions from China
X R Chen, K T Fang and C C Yang

259 Multiplication of distributions and applications to partial differential equations
M Oberguggenberger

260 Numerical analysis 1991
D F Griffiths and G A Watson

261 Schur's algorithm and several applications
M Bakonyi and T Constantinescu

262 Partial differential equations with complex analysis
H Begehr and A Jeffrey

263 Partial differential equations with real analysis
H Begehr and A Jeffrey

264 Solvability and bifurcations of nonlinear equations
P Drábek

265 Orientational averaging in mechanics of solids
A Lagzdins, V Tamuzs, G Teters and A Kregers

266 Progress in partial differential equations: elliptic and parabolic problems
C Bandle, J Bemelmans, M Chipot, M Grüter and J Saint Jean Paulin

267 Progress in partial differential equations: calculus of variations, applications
C Bandle, J Bemelmans, M Chipot, M Grüter and J Saint Jean Paulin

268 Stochastic partial differential equations and applications
G Da Prato and L Tubaro

269 Partial differential equations and related subjects
M Miranda

270 Operator algebras and topology
W B Arveson, A S Mishchenko, M Putinar, M A Rieffel and S Stratila

271 Operator algebras and operator theory
W B Arveson, A S Mishchenko, M Putinar, M A Rieffel and S Stratila

272 Ordinary and delay differential equations
J Wiener and J K Hale

273 Partial differential equations
J Wiener and J K Hale

274 Mathematical topics in fluid mechanics
J F Rodrigues and A Sequeira

275 Green functions for second order parabolic integro-differential problems
M G Garroni and J F Menaldi

276 Riemann waves and their applications
M W Kalinowski

277 Banach C(K)-modules and operators preserving disjointness
Y A Abramovich, E L Arenson and A K Kitover

Y A Abramovich
IUPUI, Indianapolis, USA
E L Arenson
Systemo-Tekhnika, St Petersburg, Russia
and
A K Kitover
Philadelphia Community College, USA

Banach $C(K)$-modules and operators preserving disjointness

Longman Scientific & Technical
Copublished in the United States with
John Wiley & Sons, Inc., New York

Longman Scientific & Technical
Longman Group UK Limited
Longman House, Burnt Mill, Harlow
Essex CM20 2JE, England
and Associated companies throughout the world.

Copublished in the United States with
John Wiley & Sons Inc., 605 Third Avenue, New York, NY 10158

First published 1992

AMS Subject Classification: (Main) 47B38, 47B60, 47A10
(Subsidiary) 47D30, 46E30, 46B42

ISSN 0269-3674

ISBN 0 582 21020 8

British Library Cataloguing in Publication Data

A catalogue record for this book is
available from the British Library

Library of Congress Cataloging-in-Publication Data

Abramovich, Y. (Yuri)
Banach C(K) modules and operators preserving disjointness / Y. Abramovich, E.L. Arenson, and A.K. Kitover.
p. cm. -- (Pitman research notes in mathematics series, ISSN 0269-3674)
ISBN 0-582-21020-8
1. Banach modules (Algebra) 2. Operator theory. 3. Banach lattices. I. Arenson, E. L. (Eugene L.) II. Kitover, A. K. (Arkady K.) III. Title. IV. Series.
QA326.A27 1992
512'.55--dc20 92-35401
CIP

Printed and bound in Great Britain
by Biddles Ltd, Guildford and King's Lynn

To our parents

Table of Contents

1. Introduction

The present work was motivated by our attempt to answer the following two basic questions:

• What are the properties, in particular the spectral properties, of (weighted) composition operators in the spaces of measurable scalar and vector valued functions?

• What classes of operators in Banach spaces are "similar" to composition operators and, at least partially, inherit their properties?

A list of works devoted to the same problems would be rather extensive. Without trying to present a complete bibliography, we mention here only several articles (see for example, [AAK, AAV1-2, Abr1-3, AL, Art, Coe1-2, Har, HP, Kam1-2, KK, Kit1-3, Lat, MaC1-2, MCS, Mon, MW, Mys, Pet, Sha, SS, ST, Wic1]). There are apparently two reasons for the continuing interest in composition operators. First, the results of the investigation of composition operators find applications in the theory of singular integral equations, dynamical systems, and differential equations with delayed time (see, for example, [KK, Mys, SS]). Secondly, the properties of composition operators are closely related to the algebraic and order structures of linear spaces, and in the investigation of these operators there arise very interesting and subtle problems, which lie on the borderlines between the theories of Banach algebras, Banach lattices, dynamical systems, and analytical functions.

Our work was inspired mostly by the second reason, but as the reader will see, many of the results presented are useful for applications as well. Among the many intertwining routes which arise in the investigation of composition and "similar" operators, we have chosen a route (still far from being fully travelled) connected with the property of *preserving disjointness.* The results obtained along this route can be divided into three seemingly isolated (but in fact closely related) groups. The division of our work into three parts corresponds to these three groups of results.

In the first part we deal with an operator from one Banach lattice into another and find conditions to guarantee that this operator may be represented as a composition operator. More precisely, we investigate the following problem: when is the obviously necessary condition of preserving disjointness also sufficient? In solving this problem we obtain a criterion for such a representation and also obtain several sufficient conditions which can be easily verified. The results in this part essentially supplement and improve the results in [Abr1, AAK].

The main focus of the second part of the work is a Banach space X and a subalgebra A of $L(X)$ (the space of all continuous linear operators) which is closed in $L(X)$ and is isometrically isomorphic to an algebra $C(K)$ for some compact Hausdorff space K. In typical situations X consists of vector valued functions and A is an algebra of scalar multipliers. The spaces with such a structure (Banach $C(K)-$modules) are "locally" similar to Banach lattices. More precisely, each closed principal A-submodule $X(x)$, generated by an arbitrary $x \in X$, has a natural structure of a Banach lattice. In terms of these local lattices $X(x)$, we are able to define several "global" concepts such as order ideals, the center of X, disjointness of elements in X, d-homomorphisms (= operators preserving disjointness), etc.

For these objects the following main results are obtained. (1) In Theorem 6.2 we give a broad generalization of a famous theorem of Bade [DS2, Ch.XVII.3]. Specifically, we prove that each linear (a priori not assumed to be continuous) operator which maps each $C(K)$-invariant subspace into itself belongs to the closure $\bar{A}$ of the algebra $A = C(K)$ in the strong operator topology of the space $L(X)$, the algebra $\bar{A}$ is a reflexive operator algebra, and $\bar{A}$ is characterized in order terms as the center of the space X. (2) The class of continuous d-homomorphisms is thoroughly studied and we give a criterion for an operator T to belong to this class. (3) On the space X^*, the conjugate of a Banach $C(K)$-module X, we define an appropriate structure of a $C(K_1)$-module and in terms of this module investigate the local order properties of X^*.

In the process, the following dual version of Bade's theorem is obtained: the center of the module X^* coincides with the space of operators for which each $\sigma(X^*, X)$-closed $C(K)$-invariant subspace of X^* is invariant. The properties of operators conjugate to d-homomorphisms are also investigated.

It is worth pointing out that most of the above mentioned results have natural analogues (and, as a rule, simpler proofs) for Banach lattices. One more comment regarding Part II is in order. There is a substantial literature devoted to the extension of Bade's theory to the framework of locally convex spaces (see for example [DP, DPR, RS, Sc, W] and the references therein). Due to some natural "boundary conditions" we do not consider this case in the present work.

The third and final part of the work is devoted to the investigation of the spectrum of d-endomorphisms (i.e., d-homomorphisms of a space into itself) of a Banach lattice or a Banach $C(K)$-module. We show that the study of the spectral properties of a d-endomorphism is closely related to the question of how the different powers of this endomorphism interact among themselves. To demonstrate this we introduce a new concept of independence of two arbitrary d-endomorphisms, a concept which in the case of Dedekind complete Banach lattices coincides with the usual disjointness of operators in the Dedekind complete vector lattice of regular operators. The main result of this part is valid for a very large class of Banach lattices and $C(K)$-modules and reads as follows: if the powers of a d-endomorphism T are pairwise independent, then the spectrum $\sigma(T)$ is rotation invariant. This theorem considerably improves all the previously known results in this direction due to Arendt, Hart, Antonevich, Lebedev, Kitover, and others. The exact references will be given in the appropriate places.

The first and the third authors would like to point out that it was E. Arenson who has inspired our investigation of $C(K)$-modules and who has contributed many original ideas.

The authors would like to express our deep gratitude to Marc Frantz and Cindy Jones for their unfailing readiness to help correct and improve our English. Our special thanks go to Mehmet Orhon for his interest and many valuable suggestions regarding the manuscript and for his advice to include in Section 9 the description of the closure of the central subalgebras in the weak* operator topology. Finally, the third author acknowledges with sincere thanks the financial aid and hospitality of the Mathematical Sciences Research Institute at Berkeley, where a part of the work was done.

2. Notations and abbreviations

For the convenience of the reader we start with a short section containing the basic notation and terminology. All the terms, symbols or definitions not explained in this section or in a corresponding place in the text are standard and may be found in [AB1], [KA], [Sch], [Vul] or [Zaa].

The following notations and abbreviations will be used throughout the work.

C (resp. **R**) denotes the field of all complex (resp. real) numbers. Unless otherwise stated explicitly, all linear spaces and vector lattices are considered over **C**.

$\mathbf{C} \cup \{\infty\}$ is the extended complex plane; we set $\lambda \cdot \infty = \infty$ for each nonzero $\lambda \in \mathbf{C}$.

$\mathbf{R} \cup \{-\infty, \infty\}$ is the extended real line.

$\mathrm{N}=\{1, 2, \ldots\}$ is the set of all natural numbers.

VL stands for an (arbitrary Archimedean) vector lattice (= Riesz space).

BL stands for a Banach lattice.

If X is a vector lattice, then $Q(X)$ denotes its Stone space. Recall that $Q(X)$ is an extremally disconnected compact Hausdorff space.

For an arbitrary extremally disconnected compact Hausdorff space Q, the symbol $C_\infty(Q)$ denotes the Dedekind complete vector lattice (and algebra) of all extended continuous functions on Q. More precisely, $C_\infty(Q)$ consists of all continuous mappings $f : Q \to \mathbf{C} \cup \{\infty\}$, such that $f^{-1}(\infty)$ is a nowhere dense subset of Q. For arbitrary $f, g \in C_\infty(Q)$ and $\lambda \in \mathbf{C}$ there exist uniquely determined functions $\lambda f, Ref, Imf, |f|, f+g$ and fg in $C_\infty(Q)$.

The real functions from $C_\infty(Q)$ will be occasionally treated as functions with values in $\mathbf{R} \cup \{-\infty, \infty\}$.

$\hat{X}$ is the Dedekind completion of a vector lattice X. Whenever appropriate, X is assumed to be embedded in $\hat{X}$.

$\mathcal{M}(X) := \mathcal{M}(\hat{X})$ is the universal completion of a vector lattice X. It is well known that when a unit e is fixed in $\mathcal{M}(\hat{X})$ the latter space can be identified with $C_\infty(Q(X))$ in such a way that e becomes the constant function one.

A **representation** j of a vector lattice X is any order isomorphic embedding $j : X \to C_\infty(Q(X))$ which agrees with the canonical isomorphism between the Boolean algebras of all bands in X and all clopen subsets of $Q(X)$.

χ_E is the characteristic function of a set E.

If X is a VL and $x_1, x_2 \in X$, then the notation $x_1 \mathrm{d} x_2$ says that the elements x_1 and x_2 are **disjoint**, i.e., $|x_1| \wedge |x_2| = 0$.

A^d is the disjoint complement of a subset A of an arbitrary vector lattice X, i.e., $A^d = \{x \in X : x \mathrm{d} a \ \forall a \in A\}$. Accordingly, $A^{dd} = (A^d)^d$.

X_x is the smallest (order) ideal containing the element x of a vector lattice X.

$X(x)$ is the smallest **norm-closed** ideal containing the element x of a Banach lattice X.

A linear operator T between two vector lattices X and Y is called a **d-homomorphism** or, equivalently, a disjointness preserving operator if $x_1 \mathrm{d} x_2$ in X implies $Tx_1 \mathrm{d} Tx_2$ in Y.

$dh(X, Y)$ denotes the collection of all regular d-homomorphisms from a vector lattice X into a vector lattice Y. Accordingly, $dh_+(X, Y) = \{T \in dh(X, Y) : T \geq 0\}$ is the subset of all lattice homomorphisms. Instead of $dh(X, X)$ we simply write $dh(X)$.

$Z(X)$ is the center of the vector lattice X, i.e., the space of all linear operators $T : X \to X$ for which there exists $\lambda = \lambda(T) > 0$ such that $|Tx| \leq \lambda |x|$ for all $x \in X$. We refer to [Wil] and [Wic2] for the standard facts about the centers. We will often use the fact that for an arbitrary Banach lattice X, its center $Z(X)$ is algebraically isomorphic to the space $C(K)$ of all continuous functions on an appropriate compact Hausdorff space K. This implies, in particular, that $Z(X)$ is a closed commutative subalgebra of the algebra $L(X)$ of all continuous linear operators on X.

Part I. Multiplicative representation

3. Multiplicative representation of d-homomorphisms on vector lattices

Let X be an arbitrary VL and $f \in Z(\hat{X})$ (hence $fx \in \hat{X}$ for each $x \in X$). Then there exists a unique function $\hat{f} \in C(Q(X))$ such that for each representation $j : \hat{X} \to C_\infty(Q(X))$ the following identity holds: $jfx = \hat{f} \cdot jx,\ x \in X$.

Allowing ourselves a bit of looseness we will identify f and $\hat{f}$.

For an arbitrary subset A of X we denote by $supp(A)$ the support set of A, i.e., the smallest clopen subset Q_A of $Q(X)$, such that $\chi_{Q_A} x = x$ for each $x \in A$. For any $x \in X$ we put $Q_x := supp(x) := supp(\{x\})$.

If Q_0 is a clopen subset of Q and $x \in C_\infty(Q_0)$, then, without any special explanations, we may (and will) consider x also as an element of the space $C_\infty(Q)$, i.e., whenever appropriate, we will identify $C_\infty(Q_0)$ with the corresponding band $\chi_{Q_0} C_\infty(Q)$ in $C_\infty(Q)$.

3.1. Definition. *Let X and Y be two vector lattices, let $T : X \to Y$ be a linear operator, and let $E = E_T = supp(TX)$. We say that T admits a multiplicative representation (m.r.) if there exist*

1) *representations $j_1 : X \to C_\infty(Q(X))$ and $j_2 : Y \to C_\infty(Q(Y))$,*

2) *a function $e \in C_\infty(Q(Y))$, and*

3) *a continuous mapping $\varphi : E \to Q(X)$,*

such that $supp(e) = E$ and

$$(j_2Tx)(q) = e(q) \cdot (j_1x)(\varphi(q)) \tag{3.1}$$

for each $x \in X$, and each $q \in Q(Y)$ satisfying $0 < |e(q)| < \infty$.

The last sentence needs some comments. Formally speaking, the function $(j_1x)\circ\varphi$ belongs to the space $C_\infty(E)$; however, in accordance with our agreement, it is treated as an element of $C_\infty(Q(Y))$. The same should be said about the function $e \cdot j_1x \circ \varphi$.

Also note that the value $(e \cdot j_1 x \circ \varphi)(q)$ of the latter function at a point $q \in E$ coincides with $e(q) \cdot (j_1 x)(\varphi(q))$ at least at each point q of the dense open subset $E_0 = \{q \in E : 0 < |e(q)| < \infty\}$ and is defined by continuity at other points. Whenever there is no danger of ambiguity we will shorten (3.1) as $j_2 T x = e \cdot j_1 x \circ \varphi$, or even as $Tx = e \cdot x \circ \varphi$ provided the representations j_1 and j_2 are fixed.

3.2. Definition. *Let X, Y and T be as above. We say that T admits a local multiplicative representation (l.m.r.) if for each $x' \in X_+$ the restriction $T|X_{x'}$, of the operator T to the ideal $X_{x'}$ generated by x', admits a m.r.*

In the case of real vector lattices it was shown in [AVK1] that *regular d-homomorphisms*, and only they, admit a l.m.r. It was also shown in [Abr1] that if X and Y are Banach lattices, then each continuous d-homomorphism is regular, and hence admits a l.m.r. On the other hand, in [Abr2] there was constructed an example of a Dedekind complete Banach lattice X and of a continuous d-homomorphism $T: X \to X$ such that T does not admit a m.r.

The purpose of this section is to generalize and/or improve the results in [Abr1] and [AAK] concerning the existence of a l.m.r. for regular d-homomorphisms, and to investigate systematically the question of the existence of a m.r. Notice that each operator admitting a l.m.r. is obviously an order bounded d-homomorphism.

Our first theorem shows that the problem of multiplicative representation for order bounded d-homomorphisms (over **C**) can be reduced to that for lattice homomorphisms. Consequently the choice of a scalar field becomes inessential, and this immediately allows us to simplify many subsequent proofs by considering the real case only.

3.3. Theorem. *Let X and Y be vector lattices, and $T: X \to Y$ a linear operator. The following conditions are equivalent.*

(1) *T is an order bounded d-homomorphism.*

(2) *$|x_1| \le |x_2|$ implies that $|Tx_1| \le |Tx_2|$.*

Each of these equivalent conditions implies that T is regular and, moreover, T admits a unique factorization $T = VT_1$ $(\star)$, where $T_1 : X \to Y$ is a lattice homomorphism, $V \in Z(\hat{Y})$ and $|V| = \chi_E$. (The support set $E = E_T$ is introduced in Definition 3.1.)

Proof. Obviously the condition $x_1 \mathrm{d} x_2$ is equivalent to the following: $|x_1| \leq |x_1 + \lambda x_2|$ for all $\lambda \in \mathbf{C}$. Hence, (2) $\Rightarrow$ (1).

To prove (1) $\Rightarrow$ (2) we can assume (since T is order bounded) that X and Y have strong units. This implies that there exist two compact Hausdorff spaces K_1 and K_2 such that X and Y may be identified with dense sublattices of $C(K_1)$ and $C(K_2)$, respectively, that $\chi_{K_1} \in X$, $\chi_{K_2} \in Y$, and that T is continuous with respect to the usual sup-norms.

Let t be an arbitrary point in K_2 such that the linear functional $l : x \mapsto (Tx)(t)$ is nonzero on X. To establish (2) it is enough to show that $|l(x_1)| \leq |l(x_2)|$ provided $|x_1| \leq |x_2|$. Let μ be a regular Borel measure in $C(K_1)^*$ which represents l and let F be the support of μ. We are done if we verify that F is a singleton. Assume by way of contradiction that F is not a singleton. Then there exist two positive functions $u_1, u_2 \in C(K_1)$ such that $u_1 \mathrm{d} u_2$ and $\int u_m \, d\mu \neq 0$ for $m = 1, 2$. Now pick $z_m \in X_+$ $(m = 1, 2)$ that are close enough (in norm) to u_m and put $x_m := (z_m - \|z_m - u_m\| \chi_{K_1})_+$. It is plain to see that $x_1, x_2 \in X$ and $x_1 \mathrm{d} x_2$, but $(Tx_1)(t) \neq 0$, $(Tx_2)(t) \neq 0$ contrary to the fact that T is a d-homomorphism. This establishes the equivalence (1) $\Longleftrightarrow$(2).

Now we will proceed to prove the validity of the factorization $(\star)$. Essentially, we have already proved (using the fact that F is a singleton) that $|T(x_1 + x_2)| = |Tx_1| + |Tx_2|$ for all $x_1, x_2 \in X_+$. This defines an additive operator $x \mapsto |Tx|$ on X_+ that can be extended in a natural way to a linear operator $T_1 : X \to Y$.

Let us fix a representation $j : Y \to C_\infty(Q(Y))$, and for each $x \in X_+$ introduce a unimodular function $V_x = jTx/|jTx|$ on Q_{Tx}. Then the identity $|jT(x_1 + x_2)| = |jTx_1| + |jTx_2|$, which is true for all $x_1, x_2 \in X_+$, obviously implies that $V_{x_1} = V_{x_2}$ on $Q_{Tx_1} \cap Q_{Tx_2}$. This and the density of the open subset $\cup\{Q_{Tx} : x \in X_+\}$ of the extremally disconnected compact space E allow us to "glue" together all the functions V_x, i.e., there exists a unique function $V \in C(Q(Y))$, such that $|V| = \chi_E$ and $V|Q_{Tx} = V_x$ for all $x \in X_+$. Thus, the factorization $T = VT_1$ is established, T_1 being a lattice homomorphism. This factorization implies immediately that T is a regular operator, and therefore the proof of Theorem 3.3 is complete. ■

Assume now that X and Y are Dedekind complete vector lattices and that a given d-homomorphism $T : X \to Y$ admits a m.r. (3.1), and let $E_T = E$, e, and φ be as in Definition 3.1. Recall that a continuous mapping $\varphi : E \to Q(X)$ generates in a canonical way an algebraic homomorphism $\gamma = \gamma_\varphi$ from $C(Q(X))$ to $C(Q(Y))$ [as always we identify $C(E)$ with the corresponding band in $C(Q(Y))$] by the formula

$$\gamma f = f \circ \varphi.$$

It is immediate from the multiplicative representation (3.1) that for all $f \in C(Q(X))$ and for all $x \in X$ the following identity holds:

$$T(f \cdot x) = \gamma(f) \cdot T(x). \tag{3.2}$$

Since $C(Q(X)) = Z(X)$ and $C(Q(Y)) = Z(Y)$ we can reformulate (3.2) as follows. If T admits a m.r., then there exists an algebraic homomorphism $\gamma : Z(X) \to Z(Y)$ such that

$$T(f \cdot x) = \gamma(f) \cdot T(x) \quad (f \in Z(X), x \in X). \tag{3.3}$$

It was an interesting and important observation due to Hart [Har], that (3.3) may be true even in the absence of (3.1). More precisely, Hart proved that if X and Y are Dedekind complete vector lattices and $T \in dh(X, Y)$, then there exists an algebraic homomorphism $\gamma_T : Z(X) \to Z(Y)$ such that (3.3) holds.† Note further that γ_T in turn generates in a standard way a continuous mapping $\varphi_T : E_T \to Q(X)$, and it is not difficult to verify that whenever T admits a m.r. (3.1), the mapping φ in (3.1) must coincide with φ_T. An alternative and independent proof of a broad generalization of Hart's result will be given in Theorem 6.7; however, for our present purposes, we need a partial generalization of this result for arbitrary vector lattices X and Y (not assumed Dedekind complete), for which the problem of the existence of φ_T is more difficult.

Let us agree (for brevity) on the following symbolism. For an arbitrary $y \in Y$, for an arbitrary clopen subset F of $Q(Y)$ containing the support set Q_y, and for an

† Notice that in view of the existence of a local m.r. for $T|X_x$ for each $x \in X$, the existence of γ_T is not that surprising. In fact, γ_T may by pieced together using these local representations, which are sufficient to recapture γ_T, but which are not sufficient to guarantee the existence of a (global) multiplicative representation.

arbitrary $g \in C(F)$, we denote by $g \cdot y$ an element $g_1 \cdot y \in \hat{Y}$, where g_1 is an arbitrary function in $C(Q(Y)) = Z(\hat{Y})$ which coincides with g on F. Obviously, this function $g \cdot y$ is independent of the choice of g_1.

3.4. Theorem. *Let X and Y be arbitrary vector lattices, $T \in dh(X,Y)$, and $E = E_T$. Then there exists a (not necessarily unique) mapping $\varphi : E \to C(Q(X))$, such that*

$$T(fx) = f \circ \varphi \cdot Tx \tag{3.4}$$

whenever $f \in Z(\hat{X}), x \in X$ and $fx \in X$. The set $\mathbf{\Phi}_T$ of all such mappings coincides with

$$\{\varphi_s : S \in dh(\hat{X},\hat{Y}),\ E_S = E \text{ and } S|X = T\}$$

Proof. To establish (3.4) we are going to use the factorization $T = VT_1$ obtained in Theorem 3.3. Notice further that in view of a theorem due to Lipecki [Lip] and Luxemburg-Schep [LS] the d-homomorphism T_1 from $dh_+(X,Y)$ can be extended to a d-homomorphism $S_1 \in dh_+(\hat{X},\hat{Y})$, in such a way that $E_{S_1} = E$. Letting $S = VS_1$, we obtain an extension of T which is a regular d-homomorphism between Dedekind complete vector lattices $\hat{X}$ and $\hat{Y}$. Consequently, the above mentioned theorem of Hart may be applied to S, and there exists therefore a corresponding mapping $\varphi = \varphi_S$. Obviously φ satisfies (3.4), since $S|X = T$.

Now, let φ be an arbitrary mapping satisfying (3.4). For each $z \in \hat{X}$ there exist $f \in Z(\hat{X})$ and $x \in X_+$ such that $z = fx$. Let us define $Sz := f \circ \varphi \cdot Tx$ and show that this expression determines a well-defined operator from $\hat{X}$ into $\hat{Y}$.

To this end it is enough to show that if $\sum_{m=1}^{\nu} f_m x_m = 0\,(*)$, where $\nu \in \mathbf{N}$, $f_m \in Z(\hat{X})$ and $x_m \in X_+$, then $\sum_{m=1}^{\nu} f_m \circ \varphi \cdot Tx_m = 0$. Let $x := \sum x_m$. Hence, $x_m = g_m x$ for an appropriate $g_m \in Z(\hat{X})$, and therefore $(*)$ becomes $(\sum f_m g_m) \cdot x = 0$. Since φ satisfies (3.4) the previous equality implies that $(\sum f_m g_m) \circ \varphi \cdot Tx = 0$, and by the same token (since $x_m = g_m x$) it implies that $\sum f_m \circ \varphi \cdot Tx_m = 0$. It is obvious that $S \in dh(\hat{X},\hat{Y})$, $S|X = T$, $E_S = E$, and $\varphi_s = \varphi$. ∎

The next corollary is just a reformulation, for further reference of our previous discussion about relationship between (3.1) and (3.4).

3.5. Corollary. *If T admits a m.r.* (3.1), *then* $\varphi \in \mathbf{\Phi}_T$.

The next result, which again is a reformulation in our notation of Proposition C of [Abr1], will be needed later on.

3.6. Theorem. *Let $T \in dh(X,Y)$ and $T = VT_1$ as in Theorem* 3.3. *Then for each $\varphi \in \mathbf{\Phi}_T$ the operator T admits a l.m.r. of the following special type. If for an $x \in X_+$ we pick up representations $j_1 : X \to C_\infty(Q(X))$ and $j_2 : Y \to C_\infty(Q(Y))$ such that $j_1(x) = \chi_{Q_x}$ and $j_2(|Tx|) = \chi_{Q_{Tx}}$, then the restriction of T to X_x admits the following representation:*

$$(j_2Tz)(q) = V(q)(j_1z)(\varphi(q)), \; z \in X_x, \; q \in Q_{Tx}.$$

3.7. Remark. (1) The set $\mathbf{\Phi}_T$ can be infinite. For example, let $X = C[0,1]$, $Y = \mathbf{C}$ and $Tx = x(0)$ $(x \in X)$. Then E_T consists of a single point q_0, and $\hat{X} = C(Q)$, where $Q = Q(X)$ is the corresponding Stone space. Each extension of T to an operator $S \in dh(\hat{X}, \mathbf{C})$ is of the form $Sz = z(p)$ $(z \in \hat{X})$, where p is an appropriate point from Q, for which $\varphi_S(q_0) = p$. Clearly, there is an infinite number of such points p.

(2) If T is order continuous, then $\mathbf{\Phi}_T$ consists of a unique mapping $\varphi = \varphi_T$ and this mapping is open. Conversely, if there exists an open mapping $\varphi \in \mathbf{\Phi}_T$, then T is order continuous. The proof consists of the verification of the following simple statements. We will assume that $T \geq 0$. Let $S_1, S_2 \in dh_+(\hat{X}, \hat{Y})$ such that $S_1|X = S_2|X = T$ and $E_{S_1} = E_{S_2} = E_T$. Then

(a) S_1 is order continuous $\iff$ φ_{S_1} is open.

(b) T is order continuous $\iff$ S_1 is order continuous.

(c) If φ_{S_1} and φ_{S_2} are open, then $S_1 = S_2$.

(3) Later we will show that if X and Y are Banach lattices and $T \in dh(X,Y)$, then operator $T^{**} \in dh(X^{**}, Y^{**})$ and T^{**} is order continuous.

Now we turn our attention to the problem of the existence of a m.r.

3.8. Theorem. *Let $T \in dh(X,Y)$, where X and Y are vector lattices. If there exists an $x \in X$ such that the band $\{Tx\}^{dd}$ generated by Tx coincides with $(TX)^{dd}$, then T admits a m.r.*

Proof. From what we have already proved it follows that, without loss of generality, we may assume that $T \geq 0$, $x \geq 0$, $\{x\}^{dd} = X$, and $\{Tx\}^{dd} = Y$. Let $\varphi \in \mathbf{\Phi}_T$. Pick up representations j_1, j_2 such that $j_1 x = \mathbf{1}$ and $j_2 Tx = \mathbf{1}$. Then (by Theorem 3.6) $j_2 Tz = (j_1 z) \circ \varphi$ for each $z \in X_x$. We want to show that the same formula is valid for each $z \in X$. Evidently we can assume additionally that $z \geq x$, since the linear combinations of such z's generate all of X. Fix such a z. Then $x = fz$ for some $f \in Z(\hat{X})$ and thus, by Theorem 3.4 , $\mathbf{1} = j_1 x = f \cdot j_1 z$ and $\mathbf{1} = j_2 Tx = f \circ \varphi \cdot j_2 Tz$. This implies that $j_1 z$ and $j_2 Tz$ are invertible elements of the algebras $C_\infty(Q(X))$ and $C_\infty(Q(Y))$, respectively. Therefore, $j_1 z = f^{-1}$ and $j_2 Tz = f^{-1} \circ \varphi = j_1 z \circ \varphi$. ∎

3.9. Remark. In both examples (constructed in [Abr1], [Abr2]) of lattice homomorphisms that do not admit a m.r., the domain vector lattice X has weak units and $(TX)^{dd} = Y$, but for no weak unit u in X is its image Tu a weak unit in Y.

Using Theorem 3.8 we can infer the following criterion for the existence of a m.r.

3.10. Theorem. *Let X and Y be vector lattices and $T \in dh(X,Y)$. The following statements are equivalent.*

(1) *T admits a m.r.*

(2) *There exists an order dense ideal X_1 in $\mathcal{M}(\hat{X})$ containing X and an operator $T_1 \in dh(X_1, \mathcal{M}(\hat{Y}))$, such that the restriction $T_1|X = T$ and T_1 satisfies the condition of Theorem* 3.8.

Proof. Obviously (2) $\Rightarrow$ (1). Conversely, assume (1), i.e., that formula (3.1) is valid for T. Without loss of generality we may assume that $(TX)^{dd} = Y$, that as usual X is embedded into $C_\infty(Q(X))$, Y is embedded into $C_\infty(Q(Y))$ and that $Tx = e \cdot x \circ \varphi$ for all $x \in X$. Now if we put $X_1 := \hat{X} + C(Q(X))$ and define an operator $T_1 : X_1 \to C_\infty(Q(Y))$ by the formula $T_1 z = e \cdot z \circ \varphi$, then we can easily see that X_1 and T_1 satisfy (2).

For Dedekind complete vector lattices the previous criterion can be reformulated in "internal" terms. To do so we recall the following definition first introduced in [Abr3].

3.11. Definition. *Let X be a Dedekind complete VL and L be a subset of X_+. We will call L* **laterally directed** *if for each $x_1, x_2 \in L$ there exists an $x \in L$ such that both x_1 and x_2 are components of x, that is,*

$$|x - x_i| \wedge |x_i| = 0 \ (i = 1, 2).$$

It is easy to see that L is an increasing directed set (possibly with a largest element) and for each $x_1 \leq x_2$ from L the element x_1 is a component of x_2.

3.12. Theorem. *Let X and Y be Dedekind complete vector lattices and T belongs to $dh(X, Y)$. The following statements are equivalent.*

(1) *T admits a m.r.*

(2) *There is a laterally directed subset L of X_+ such that $(TL)^{dd} = (TX)^{dd}$.*

Proof. (1) $\Rightarrow$ (2). Let T admit a m.r. (3.1). Using this representation we can define the following set

$$L = \{x \in X_+ : j_1 x \text{ is a component of } \mathbf{1} = \chi_{Q(X)}\}.$$

It is plain to see that L is laterally directed and $(TL)^{dd} = (TX)^{dd}$, i.e., condition (2) holds.

Conversely, assume condition (2). Without loss of generality we can assume that the operator $T \geq 0$ and that $(TX)^{dd} = Y$. Moreover, we can assume for simplicity that $L^{dd} = X$. Lateral directedness of L implies (see [Vul], Lemma V.2.1 and a footnote to it) that in $\mathcal{M}(\hat{X})$ there exists $x_0 = \sup L$, and similarly in $\mathcal{M}(\hat{Y})$ there exists $y_0 = \sup TL$. Obviously x_0 and y_0 are weak units in $\mathcal{M}(\hat{X})$ and $\mathcal{M}(\hat{Y})$ respectively. Pick representations j_1 and j_2 such that $j_1 x_0 = \chi_{Q(X)}$ and $j_2 y_0 = \chi_{Q(Y)}$. We will show now that for each $x \in X$ we have $j_2 T x = (j_1 x) \circ \varphi$, where $\varphi = \varphi_T$. For simplicity of notation we identify X with $j_1 X$ and Y with $j_2 Y$. Then $L \subset C(Q(X))$ and L consists of the components of the unit in $C(Q(X))$. For each $z \in L$ the element Tz is a component of the unit in $C(Q(Y))$.

In view of Theorem 3.6, $Tz = z \circ \varphi$ for each $z \in L$. Let $x \in X \cap C(Q(X))$. Then applying the equality (3.4) twice (first to z as to an element of $Z(\hat{X})$, and after that similarly to x) we have

$$z \circ \varphi \cdot Tx = T(zx) = x \circ \varphi \cdot Tz = z \circ \varphi \cdot x \circ \varphi.$$

Since z is arbitrary and since $(TL)^{dd} = (TX)^{dd}$ the above identity implies that $Tx = x \circ \varphi$ for all x under consideration. To complete the proof let us introduce the vector lattice $X_1 := X + C(Q(X))$, define $T_1 : X_1 \to C_\infty(Q(Y))$ by the formula $T_1(x + f) = Tx + f \circ \varphi$ $(x \in X, f \in C_\infty(Q(X))$, and then apply Theorem 3.10. ■

3.13. Corollary. *Let X and Y be vector lattices and $T \in dh(X, Y)$. If there exists in X a subset M of pairwise disjoint elements such that $(TM)^{dd} = (TX)^{dd}$, then T admits a m.r.*

Proof. It is enough to notice that the condition of the corollary remains valid when we extend T to an operator $\hat{T} \in dh(\hat{X}, \hat{Y})$, and that the set M_1 of finite sums of absolute values of elements of M is laterally directed. After that one can apply the previous theorem to $\hat{T}$ and M_1. ■

3.14. Remark. A. V. Koldunov has constructed an interesting example of Dedekind complete vector lattices X and Y, and an operator $T \in dh_+(X, Y)$ such that T admits a m.r., but for each subset $M \subset X$ of pairwise disjoint elements, $(TM)^{dd} \neq (TX)^{dd}$.

Using Theorem 3.12 we can describe a class of "good" vector lattices, for which each regular d-homomorphism admits a m.r.

3.15. Theorem. *Let X be a Dedekind complete BL. The following two conditions are equivalent.*

(1) *For an arbitrary vector lattice Y any operator $T \in dh(X, Y)$ admits a m.r.*

(2) *There exists a laterally directed set $L \subset X_+$ such that the norm closed order ideal $X(L)$, generated by L, coincides with X.*

Proof. (1) $\Rightarrow$ (2). We denote by $\mathcal{H}$ the collection of all norm-closed proper ideals in X. For each $H \in \mathcal{H}$ the canonical quotient mapping $T_H : X \to X/H$ is a lattice homomorphism. Let us introduce $Y := (\sum_{H\in\mathcal{H}} \oplus X/H)_{\ell_\infty}$, the direct ℓ_∞-sum, and define $T : X \to Y$ by $Tx = \{T_H x\}_{H\in\mathcal{H}}$, $(x \in X)$. Obviously $T \in dh(X,Y)$, and, by Theorem 3.12, condition (1) implies that there exists a laterally directed subset $L \in X_+$, such that $(TL)^{dd} = Y$. If $X(L) \neq X$, then the ideal $H = X(L)$ would belong to $\mathcal{H}$ and for each $x \in L$ we would have $T_H x = 0$. But then $(TL)^{dd} \neq Y$. A contradiction. Hence (1)⇒(2).

Conversely, assume (2). Then $X(L) = X$ for some laterally directed $L \in X_+$. Fix an arbitrary vector lattice Y and an arbitrary $T \in dh(X,Y)$. We denote by P_T a band projection in $\hat{Y}$ onto the band $(TL)^d$ and let

$$I_T = \{x \in X : \ P_T Tx = 0\}.$$

Then obviously I_T is an ideal in X and $L \subset I_T$. It is easy to verify that I_T is closed in X in the relative-uniform topology (r_u-topology), and consequently, I_T is norm-closed ([Vul], Thm.VII.2.1). Hence $I_T = X$ (since $X(L) = X$), and this means that $(TL)^{dd} = (TX)^{dd}$. Theorem 3.12 implies now that T admits a m.r. ■

3.16. Remark. In a similar way one can prove a generalized version of the previous theorem for an arbitrary Dedekind complete vector lattice X, provided one defines $X(L)$ to be the smallest (r_u)-closed ideal generated by L.

The previous results can be applied to infer several useful and easily verified sufficient conditions.

3.17. Theorem. *Each of the following conditions implies that $T \in dh(X,Y)$ admits a m.r.*

(1) *T is order continuous.*

(2) *X is a Banach lattice with order continuous norm.*

(3) *X is a separable Banach lattice.*

(4) *The Banach lattice X has a quasi-interior point x_0, i.e., there is an element $x_0 \in X_+$ such that $X(x_0) = X$.*

(5) *Every subset of pairwise disjoint non-zero elements in Y is at most countable, or equivalently, $\mathcal{M}(\hat{Y})$ satisfies the countable sup property.*

Proof. Obviously (2) $\Rightarrow$ (1) and (3) $\Rightarrow$ (4). The sufficiency of (1) follows from Remark 3.7(2) and Corollary 3.13. (If $\varphi = \varphi_T$ is open, then the equality $(TM)^{dd} = (TX)^{dd}$ holds for each subset M of pairwise disjoint elements of X satisfying $M^{dd} = X$.) The sufficiency of (4) is proved in [Abr2] and it can also be immediately inferred from Theorem 3.15. Indeed, each quasi-interior point $x_0 \in X_+$ is also a quasi-interior point in $\hat{X}$ endowed with the natural norm $\|z\|_{\hat{X}} = \inf\{\|x\| : x \in X_+, \ |z| \le x\}$. Setting $L = \{x_0\}$ we obtain $\hat{X}(L) = \hat{X}$.

To finish the proof of the theorem we need to show that condition (5) is also sufficient. First we notice that this condition still holds if we replace T by its extension $\hat{T} \in dh(\hat{X}, \hat{Y})$. Thus, we can assume that X and Y are Dedekind complete vector lattices. Second we notice that (5) implies that there is a countable subset $A = \{x_n\}_{n=1}^{\infty} \subset X$ such that $(TA)^{dd} = (TX)^{dd}$. Let us define by induction $z_1 = x_1, \ldots, z_{n+1} = (P_{n+1} - P_n)x_{n+1}$, where P_n is the band projection on the band generated by the element $|x_1| \vee \ldots \vee |x_n|$. Then the set $M = \{z_n\}_{n=1}^{\infty}$ satisfies the condition of Corollary 3.13, and thus T admits a m.r. ■

Recall that in [Abr2] in an example of an operator $T \in dh(X, Y)$ that does not admit a m.r., X is a nonseparable Orlicz space. Now, in view of Theorem 3.17(3), we see that this assumption is indispensable. Moreover, the next theorem shows that for a large class of Banach lattices, namely for the rearrangement invariant Banach function spaces (see their definition in [LT]), this assumption is also necessary.

3.18. Theorem. *Let X be a rearrangement invariant Banach function space on* [0,1]. *The following conditions are equivalent.*

(1) *The function* **1** *is a quasi-interior element in X (or, equivalently, $L_\infty[0,1]$ is norm dense in X).*

(2) *For an arbitrary vector lattice Y any operator $T \in dh(X, Y)$ admits a m.r.*

Moreover, if $X \neq L_\infty[0,1]$, then each of the previous conditions is equivalent to the following one:

(3) *X is separable.*

Proof. The validity of (1) $\Rightarrow$ (2) follows from Theorems 3.17(4) or 3.15. To prove that (2) $\Rightarrow$ (1) we first note that by Theorem 3.15 $X(L) = X$ for some laterally directed $L \subset X_+$. Without loss of generality we can assume that for each $x \in L$ each of its components also belongs to L. Then there exists an $x \in L$ and a number $a > 0$ such that the set $F = supp(x)$ satisfies the following two conditions: the measure of F equals 1/2 and $a\chi_F \leq x \leq a^{-1}\chi_F$. Let us set $X_1 := \chi_F X$ and $L_1 := \chi_F L$. The condition $X(L) = X$ obviously implies that $X(L_1) = X_1$. Since $x = \sup L_1$, we have $X_1(x) = X_1$ or, equivalently, $X_1(\chi_F) = X_1$. Therefore χ_F is a quasi-interior point in X_1. Let $G = Q(X) \setminus F$. The rearrangement invariance of X clearly implies that χ_G is a quasi-interior point of $X_2 = \chi_G X$. This implies that $\mathbf{1} = \chi_F + \chi_G$ is a quasi-interior point of X.

Finally, let $X \neq L_\infty[0,1]$. We need to prove only the implication (1) $\Rightarrow$ (3), since (3) $\Rightarrow$ (2) holds by the previous theorem. It is well known that the restriction of the invariant norm $\|\cdot\|_X$ to $L_\infty[0,1]$ is order continuous, and thus $(L_\infty[0,1], \|\cdot\|_X)$ is separable. But in view of (1) the space $L_\infty[0,1]$ is dense in X. Therefore, X is separable. ■

3.19. Remark. Suppose $T \in dh(X,Y)$ admits a m.r. (3.1).

(1) If j_1 and φ are fixed, then formula (3.1) remains valid for an arbitrary replacement of j_2 by any other representation j_2' (with a corresponding replacement of the weight e), and for an arbitrary replacement of e by an arbitrary function e' with the same support as e (with a corresponding replacement of the representation j_2). In particular, (a) there exists a m.r. (3.1) for which $|e| = \chi_E$, and (b) if $X = Y$, then we can replace j_2 by j_1 and thus T admits a m.r. with a single representation $j = j_1 = j_2$. This possibility is essential when one investigates the spectral properties of d-homomorphisms.

(2) A replacement of the representation j_1 by an arbitrary representation is not possible in general. However, if T is order continuous, then T admits a m.r. for arbitrary representations j_1 and j_2.

(3) If X and Y are two order dense ideals in a vector lattice W, then there exist two representations j_1' and j_2' of W, such that (for some mapping φ and some function e) formula (3.1) is valid for $j_1 = j_1'|X$ and $j_2 = j_2'|Y$.

(4) Let X be an arbitrary vector lattice and $T \in Z(X)$ be a central operator, i.e.,

$|Tx| \leq \lambda|x|$ $(*)$ for some $\lambda \geq 0$ and all $x \in X$. It is obvious that $(*)$ implies that T is order continuous, and consequently, in view of Theorem 3.17(1) and Remark (1) above, T admits a m.r. of the form $Tx = e \cdot x \circ \varphi$. Two other immediate consequences of $(*)$ are that $\varphi(q) = q$ for all $q \in E_T$ and that e belongs to $C(Q)$ (rather than to $C_\infty(Q)$). That is, $Tx = e \cdot x$, and this means that T is a multiplication (by a complex-valued continuous function), not only on the appropriate compact space mentioned at the very end of Section 2, but also on the Stone space $Q(X)$. Another proof of this result may be obtained by extending $T \in Z(X)$ to an operator $\hat{T} \in Z(\hat{X})$.

Part II. C(K)-modules and d-homomorpisms

4. Banach C(K)-modules and their relationship with Banach and vector lattices

Let K be a compact Hausdorff space. We say that a Banach space X is a **Banach C(K)-module** (or, shorter, a **C(K)-module**) if there is a bilinear mapping $(f, x) \mapsto f \cdot x$ from $C(K) \times X$ into X satisfying the following conditions:

$$\mathbf{1} \cdot x = x,\ (fg) \cdot x = f \cdot (g \cdot x) \quad \text{and} \quad ||f \cdot x|| \leq ||f||_{C(K)}||x||$$

for each $f, g \in C(K)$ and $x \in X$. †

Usually, when this does not cause any ambiguity, we write simply fx instead of $f \cdot x$. Numerous examples of $C(K)$-modules will be considered in Section 14.

One of our goals is to show that there is a very deep parallelism between the properties of $C(K)$-modules on one hand and the properties of vector and Banach lattices on the other. Since there are many examples of $C(K)$-modules which are not Banach lattices, this parallelism allows us to develop for $C(K)$-modules a deep theory reflecting the one existing for Banach lattices. Moreover, the relationships between theories work both ways. Some special properties of $C(K)$-modules allow us to look differently at Banach lattices and to deepen their theory, too.

Recall that if X is a vector lattice, then for an arbitrary $x \in X$, the (order) interval‡ generated by x is the following set

$$\Delta(x) = \{y \in X : |y| \leq |x|\,\}.$$

† It is worth mentioning that a more general case which assumes only the inequality $||f \cdot x|| \leq \gamma ||f||_{C(K)}||x||$ for some $\gamma \geq 1$ can be reduced to the case under consideration with $\gamma = 1$ by an equivalent renorming of X.

‡ This term is standard in the case of real vector lattices, and we preserve it for the complex case, too.

The next definition establishes a natural bridge between Banach lattices and $C(K)$-modules by postulating which objects in $C(K)$-modules should serve as a correct counterpart of intervals in Banach lattices. We preserve for these objects the same symbol $\Delta(x)$, though occasionally (when some ambiguity is possible because of the different algebras involved) we will use also symbol $\Delta_{C(K)}(x)$ to express explicitly the algebra generating this interval.

4.1. Definition. *If X is an arbitrary $C(K)$-module and $x \in X$, then*

$$\Delta(x) = \mathrm{cl}_X\{fx : f \in C(K), ||f|| \leq 1\},$$

where "cl_X" denotes the norm closure in X.

In the next few definitions we are going to use this set-valued mapping $x \mapsto \Delta(x)$ to extend to $C(K)$-modules the main concepts which were previously known (and used) in the context of vector and Banach lattices only.

4.2. Definition. *A linear subspace Y of a $C(K)$-module X is said to be an (order) ideal if for each $x \in Y$ the whole interval $\Delta(x)$ belongs to Y.*

It is easy to see that each ideal of a $C(K)$-module is a $C(K)$-submodule, not necessarily closed. The converse statement is not true in general; but each closed $C(K)$-submodule is an ideal. It is also clear that the closure of each ideal is likewise an ideal.

4.3. Definition. *Let Y be an arbitrary ideal in a $C(K)$-module X. We denote by $Z_{C(K)}(Y)$, or simply by $Z(Y)$, the collection of all linear operators $T : Y \to Y$, such that $Ty \in \lambda\Delta(y)$ for each $y \in Y$, where $\lambda = \lambda(T)$ is a positive number (independent of y), i.e.,*

$$Z(Y) = \{T : Y \to Y \mid (\exists\lambda \geq 0)(\forall y \in Y)(Ty \in \lambda\Delta(y))\}.$$

Obviously, $Z(Y)$ is an algebra of operators on Y. We call this algebra the **center** *of Y.*

4.3.1. Remark. In Theorem 6.2, one of the main results of Section 6, we will show that the assumption that λ is independent of the element y is actually redundant, provided T is an operator on the whole of a $C(K)$-module X.

4.3.2. Remark. It is obvious that for each $x' \in \Delta(x)$ we have $\Delta(x') \subseteq \Delta(x)$. Therefore, for each $T \in Z(X)$ we have $Tx' \in \lambda\Delta(x') \subseteq \lambda\Delta(x)$, i.e., $T(\Delta(x)) \subseteq \lambda\Delta(x)$.

4.4. Definition. *Two elements x, y of a $C(K)$-module X are called disjoint (in symbols: $x\mathrm{d}y$), if $\Delta(x+y) = \Delta(x) + \Delta(y)$ and $\Delta(x) \cap \Delta(y) = \{0\}$.*

4.4.1. Remark. The subsequent Lemma 4.8(4) shows that our definition of disjointness in $C(K)$-modules is adequate for our pursuit of the parallelism between Banach lattices and $C(K)$-modules. If X is a Banach lattice, then evidently $x\mathrm{d}y \iff \Delta(x) \cap \Delta(y) = \{0\}$ and each of these implies that $\Delta(x+y) = \Delta(x) + \Delta(y)$. However, the last relation $\Delta(x+y) = \Delta(x) + \Delta(y)$ alone does not imply that $x\mathrm{d}y$, because $\Delta(2x) = \Delta(x) + \Delta(x)$ for each $x \in X$.

It seems somewhat curious that if X is a $C(K)$-module then, in general, the condition $\Delta(x)\cap\Delta(y) = \{0\}$ does not imply that $\Delta(x+y) = \Delta(x)+\Delta(y)$. For example, let X be the $L^\infty[0,1]$-module $L^2[0,1] \oplus L^2[0,1]$, and let $x = (\mathbf{1}, 0)$ and $y = (0, \mathbf{1})$. Then $\Delta(x+y) \neq \Delta(x) + \Delta(y)$.

We leave it as an exercise for the reader to prove that whenever f_1 and f_2 are arbitrary disjoint functions in $C(K)$, the elements f_1x_1 and f_2x_2 are disjoint in the $C(K)$-module X for arbitrary $x_1, x_2 \in X$.

4.5. Definition. *Let $T : X \to Y$ be a linear operator from a $C(K_1)$-module X into a $C(K_2)$-module Y. We call operator T a d-homomorphism (a d-endomorphism provided $X = Y$) if $x_1\mathrm{d}x_2$ in X implies $Tx_1\mathrm{d}Tx_2$ in Y.*

An invertible operator $T : X \to Y$ is called a d-isomorphism if T and T^{-1} are d-homomorphisms.

We want to emphasize once again that for vector lattices the above definitions define the conventional objects carrying the same names.

Observe that in view of Remark 4.3.2 and Lemma 4.8(4) each operator T from $Z(X)$, where X is an arbitrary $C(K)$-module, is a d-endomorphism.

4.5.1. If T is a continuous d-endomorphism on a Banach lattice X and T is invertible, then T^{-1} is likewise a d-endomorphism. However, it is curious to notice that if X is a $C(K)$-module and T is an invertible continuous d-endomorphism on X, then T^{-1} may not be a d-endomorphism. A corresponding example will be given in Remark 12.17.

4.5.2. Throughout the work, we will use the following notations, which are also in complete agreement with their vector lattice counterparts. If G is a subset of a $C(K)$-module X, then X_G denotes the order ideal generated by G, and $X(G)$ denotes the closure of X_G in X. In case $G = \{x\}$ is a singleton, we prefer a simpler notation X_x and $X(x)$, instead of $X_{\{x\}}$ and $X(\{x\})$, respectively. It is plain to see that the centers $Z(X_x)$ and $Z(X(x))$ are canonically isomorphic and, therefore, we will denote them simply by $Z(x)$.

If X is a Banach lattice, then the same identification between $Z(X_x)$ and $Z(X(x))$ will be always assumed and, again, the same symbol $Z(x)$ will be used to denote either of them. However, in this case the objects X_x, $X(x)$ and their respective centers should, of course, be understood in the sense appropriate for Banach lattices, that is, as they are defined in the introduction.

4.6. Lemma. *Let x be an arbitrary element of a $C(K)$-module X. Let*

$$X(x)_+ = \mathrm{cl}_X\{fx : 0 \le f \in C(K)\}.$$

Then the space $X(x)$ ordered by the cone $X(x)_+$ is a Banach lattice. Moreover, the interval mapping $x' \mapsto \Delta(x')$ on this Banach lattice $X(x)$ coincides with the mapping $x' \mapsto \Delta_{C(K)}(x')$ generated in $X(x)$ by the structure of the $C(K)$-module. (That is why it does not cause any ambiguity to use the same symbol for both mappings.)

Proof. If X is a cyclic $C(K)$-module, which means that there exists an element $x_0 \in X$ such that $\mathrm{cl}\{fx_0 : f \in C(K)\} = X$, then the first statement was proved by Kaijser [Kai]. The case of general $C(K)$-modules can be routinely reduced to the previous one. The second statement of the lemma follows easily from the definitions. ∎

4.6.1. If X is an arbitrary Banach lattice and $x \in X$, then the closed ideal $X(x)$, generated by x, and with the order induced from X, is a Banach lattice, a sublattice of X. On the other hand, since its center $Z(x)$ can be identified with a $C(K)$-space, where K is the maximal ideal space of the algebra $Z(x)$, the space $X(x)$ is as well a $C(K)$-module, and consequently, by Lemma 4.6, the structure of this $C(K)$-module turns $X(x)$ into another Banach lattice. Though, in general, the orders of these Banach lattices are different (they coincide iff $x \in X_+$), the mappings $x \mapsto \Delta(x)$ generated by these orders coincide, since (as collections of elements) $X(x) = X(|x|)$ and $Z(x) = Z(|x|)$.

From now on, when speaking about the order on $X(x)$ (be X a Banach lattice or a $C(K)$-module), we will always understand (unless otherwise stated explicitly) that $X(x)$ is viewed as a Banach lattice in accordance with Lemma 4.6.

4.6.2. Let X be a $C(K)$-module, Y be an ideal in X and $T \in Z(Y)$. For each $x \in Y$ we let $\lambda(T, x) = \inf\{\lambda > 0 : Tx \in \lambda\Delta(x)\}$. Since x is a quasi-interior point in the Banach lattice $X(x)$ and $T|X_x \in Z(x)$ we have $\|T|X_x\| = \|T|X(x)\| = \lambda(T, x)$. Therefore $\|T\| = \sup_{x \in Y} \|T|X_x\| = \sup_{x \in Y} \lambda(T, x)$. This implies that all operators in $Z(Y)$ are continuous and that the set $\{T \in Z(Y) : \|T\| \leq 1\}$ is closed in the strong operator topology. In particular, $Z(Y)$ is a closed subalgebra of the algebra $L(X)$.

Let us notice also that there obviously exists a natural algebraic homomorphism of $C(K)$ into $Z(X)$. This homomorphism does not increase the norm and, in general, is neither injective nor surjective.

The next technical definition plays an important role throughout the whole work.

4.7. Definition. *Suppose $x \in X$ and $x^* \in X^*$ are two fixed elements, where X is a $C(K)$-module. Then we denote by $x^* \square\, x$ the following functional from $C(K)^*$ (that is, a regular Borel measure on K)*

$$(x^* \square\, x)(f) = x^*(fx), \quad f \in C(K).$$

In other words, we have a bilinear mapping $(x^*, x) \mapsto x^* \square\, x$ with the values in the AL-space $C(K)^*$. It can be readily seen that $\|x^* \square\, x\| \leq \|x^*\| \cdot \|x\|$.

4.8. Lemma. *Let X be a $C(K)$-module and $x, y \in X$. The following statements are true.*

1) *A functional $x^* \in X^*$ is orthogonal to $X(x) \iff x^* \square\, x = 0$.*

2) *A functional $x^* \in X^*$ is nonnegative on $X(x)_+ \iff x^* \square\, x \geq 0$ as an element of $C(K)^*$.*

3) *$y \in \Delta(x) \iff \|x^* \square\, y\| \leq \|x^* \square\, x\|$ for each $x^* \in X^* \iff |x^* \square\, y| \leq |x^* \square\, x|$ for each $x^* \in X^*$.*

4) *$x \mathrm{d} y \iff$ the elements x and y belong to the vector lattice $X(x+y)$ and are disjoint in it $\iff (x^* \square\, x)\mathrm{d}(x^* \square\, y)$ in $C(K)^*$ for each $x^* \in X^*$.*

Proof. Statements 1) and 2) follow easily from definitions. Let A, B and C (respectively) denote three conditions which are claimed to be equivalent in 3). Implication $A \Longrightarrow C$ follows from the definition of the set $\Delta(x)$; implication $C \Longrightarrow B$ is evident; and $B \Longrightarrow A$ follows by a standard separation theorem.

Let now D, E and F (respectively) denote the three conditions, the equivalence of which is claimed in 4). The equivalence $D \iff E$ can be inferred easily from the definition of disjointness in a $C(K)$-module. The equivalence $E \iff F$ follows from 3). ∎

4.9. Corollary. *Let X be a Banach lattice or a $C(K)$-module. Then the graph $\{(x,y) : x \in X,\ y \in \Delta(x)\}$ of the mapping $x \longrightarrow \Delta(x)$ is closed in $X \times X$.*

Proof. For a Banach lattice the statement is obvious. For a $C(K)$-module it follows from Lemma 4.8(3). ∎

4.10. Corollary. *Let Y be a $C(K)$-submodule of a $C(K)$-module X, and let Y_1 denote the ideal in X generated by Y. Assume also that $U : Y \to X$ is a linear operator such that $Uy \in \lambda\Delta(y)$ for each $y \in Y$, where $\lambda \geq 0$ is a fixed positive number. Then U admits a unique extension $V \in Z(Y_1)$.*

Proof. Obviously U is continuous, and a fortiori, it has a unique continuous extension V_1 on $X_1 = \mathrm{cl}_X Y$. It is readily seen that X_1 is again a $C(K)$-submodule of X, and, being closed, it is an ideal in X. Therefore $Y_1 \subseteq X_1$. By Corollary 4.9, $V_1 \in Z(X_1)$, and hence $V_1(Y_1) \subset Y_1$. It remains to set $V := V_1|_{Y_1}$. ■

The next definition introduces an object which is of crucial importance for what will follow.

4.11. Definition. *If X is a $C(K)$-module, then L^1_X denotes the band in $C(K)^*$ generated by all measures $x^* \square x$, where $x^* \in X^*$, and $x \in X$. Consequently L^1_X is a Banach lattice, and moreover, an AL-space.*

4.11.1. Lemma. *L^1_X coincides with the closed subspace of $C(K)^*$ generated by all measures $x^* \square x$, where $x^* \in X^*$, and $x \in X$.*

Proof. It is enough to show that if $\nu \in C(K)^*$ satisfies $|\nu| \leq |x^* \square x|$ for some $x^* \in X^*$ and $x \in X$, then $\nu = y^* \square x$ for some $y^* \in X^*$. Let $\mu = x^* \square x$. By the Radon-Nikodym theorem $\nu = h\mu$ for some $h \in L^\infty(|\mu|)$ with $\|h\|_\infty \leq 1$. Take any sequence $\{f_n\}$ in the unit ball of the space $C(K)$ which converges to h in $L^1(|\mu|)$-norm. For each $n \in \mathbf{N}$ we denote by y^*_n a functional in X^* defined as follows: $y^*_n(z) = x^*(f_n z)$, $z \in X$. Clearly $\|y^*_n\| \leq \|x^*\|$. Let y^* be an arbitrary $\sigma(X^*, X)$-limit point of $\{y^*_n\}$.

Then for each $f \in C(K)$ we have

$$\int f\,d\nu = \int fh\,d(x^* \square x) = \lim_n \int ff_n\,d(x^* \square x) = \lim_n x^*(f_n fx) =$$

$$= \lim_n y^*_n(fx) = y^*(fx) = \int f\,d(y^* \square x),$$

whence $\nu = y^* \square x$. ■

4.12. Recall that a subset D of a vector lattice V is called *solid* if $(|v| \leq |x|, x \in D, v \in V)$ implies that $v \in D$.

It is important to notice that in the course of the proof of Lemma 4.11.1 we showed that the set $D = \{x^* \square x : x^* \in X^*, x \in X\}$ of all "elementary" measures is a solid subset of L^1_X.

In addition we will need the following approximation result, whose simple proof is omitted.

4.12.1. Lemma. *Let D be a solid subset of a Dedekind complete vector lattice V. Then for each v in V_D, the order ideal generated by D, there are pairwise disjoint elements $\{x_k\}_{k=1}^n \subset D$ and numbers $\{\lambda_k\}_{k=1}^n \subset \mathbf{C}$ such that $v = \Sigma \lambda_k x_k$.*

4.12.2. Remark. The assumption that V is Dedekind complete is essential.

4.12.3. Corollary. *For each $\mu \in L^1_X$ and for each $\epsilon > 0$ there exists a finite collection of pairwise disjoint measures $x_k^* \square x_k$, $x_k^* \in X^*, x_k \in X$, and numbers $\lambda_1, \ldots, \lambda_n \in \mathbf{C}$, such that $||\mu - \Sigma \lambda_k x_k^* \square x_k|| \leq \epsilon$.*

4.12.4. Corollary. *For each $\mu \in L^1_X$ there exists a sequence of pairwise disjoint measures $x_k^* \square x_k$, $x_k^* \in X^*, x_k \in X$, such that $\mu \in \{x_k^* \square x_k\}^{dd}$, i.e., the band generated by this sequence contains μ.*

Remark. Later on, in Theorem 8.1, we will be able to prove that in actuality $L^1_X = \{x^* \square x : x^* \in X^*, x \in X\}$, which is a much stronger result than the preceding two corollaries and Lemma 4.11.1.

4.13. Another structure on L^1_X which can be introduced immediately is that of a $C(K)$-module. Indeed, for each $f \in C(K)$ and each measure $\mu \in L^1_X$, the product $f \cdot \mu$ is defined as the following functional on $C(K)$:

$$(f \cdot \mu)(g) = <\mu, fg> = \mu(fg) \ (g \in C(K)).$$

It is plain to verify that $f \cdot \mu \in L^1_X$, and that under this definition L^1_X becomes a $C(K)$-module.

4.13.1. For further references note that the above introduced multiplication of a function $f \in C(K)$ by a measure $\mu = x^* \square x \in L^1_X$, $x^* \in X^*, x \in X$, enjoys the following useful property:

$$f \cdot (x^* \square x) = x^* \square fx. \tag{4.1}$$

Indeed, for an arbitrary $g \in C(K)$ we have:

$$<f \cdot (x^* \square x), g> = <x^* \square x, fg> = <x^* \square x, gf> =$$

$$<x^*, (gf)(x)> = x^*(g(fx)) = <x^* \square fx, g>.$$

The first equality follows from the definition of the multiplication, the second from commutativity in $C(K)$, the third and the last from the definition of the operation $\square$, and the second to last follows from associativity. Since g is arbitrary, this proves (4.1). Note also that $x^* \square fx = (f^*x^*) \square x$ because $< (f^*x^*) \square x, g >= (f^*x^*)(gx) = x^*(f(gx))$ for arbitrary $g \in C(K)$.

4.13.2. Finally, denote by L_X^∞ the center $Z(L_X^1)$ of the Banach lattice L_X^1. Since L_X^1 ia an AL-space, the algebra L_X^∞ is an AM-space, which is isometric and order isomorphic to the conjugate space $(L_X^1)^*$. Therefore, L_X^1 may be viewed also as an L_X^∞-module.

Using the above notation we can easily prove the following proposition.

4.14. Proposition. *The mappings $\mu \longrightarrow \Delta(\mu)$, determined on L_X^1 by the structures of a Banach lattice, of a $C(K)$-module, and of an L_X^∞-module coincide. That is, $\Delta(\mu) = \Delta_{C(K)}(\mu) = \Delta_{L_X^\infty}(\mu)$.*

4.15. Let A be an ideal in a $C(K)$-module X. Since A is a $C(K)$-module in its own right we can consider the space L_A^1 which is a band in $C(K)^*$. It is important to observe that L_A^1 is actually contained in L_X^1, and therefore L_A^1 is a band in L_X^1. This observation follows from the following simple fact. For arbitrary $x^* \in X^*$ and $a \in A$, the measures $x^* \square a \in L_X^1$ and $(x^*|A) \square a \in L_A^1$ coincide.

5. Comparison of operators taking values in a vector lattice or C(K)-module

Throughout this section X stands for a vector lattice or a $C(K)$-module, n is a natural number; $E_1, E_2 \ldots, E_n$ are linear spaces, and $E = E_1 \times E_2 \cdots \times E_n$ denotes their cross-product; $T : E \to X$ is an n-linear operator; and finally, $X_1 = X_{T(E)}$ is the ideal in X generated by the set $T(E)$.

5.1. Definition. *We denote by the symbol $Z \circ T$ the collection of all n-linear operators $S : E \to X$ admitting a representation $S = VT$ for some $V \in Z(X_1)$.*

Since V maps X_1 into X_1, we can consider V as a mapping from X_1 into X, and therefore, the operator S is well-defined.

If $E = X$ and $T = \mathrm{id}_X$, then $Z \circ T = Z(X)$, and thus, the set $Z \circ T$ of operators may be viewed as a generalization of the center.

The next lemma follows easily from the previous definition, and its proof is omitted.

5.2. Lemma. *Let $S \in Z \circ T$. The following statements are true.*

(1) *The factorization $S = VT$,* with *$V \in Z(X_1)$, is unique.*

(2) *If X is a Dedekind complete VL, then there exists a unique factorization $S = WT$ with $W \in Z((TE)^{dd})$, or, what is essentially the same, $W \in Z(X)$ and $W \equiv 0$ on $(TE)^d$.*

The goal of this section is to present several easily verifiable sufficient conditions for S to belong to $Z \circ T$.

5.3. Definition. *We say that an operator $S : E \to X$ is* **dominated** *by T on the elements of E if for each $e \in E$ there is a constant $c \geq 0$ such that $Se \in c\Delta(Te)$. In this case, we use the notation $S \prec T$. If such a constant $c > 0$ may be chosen independently of $e \in E$, then we say that S is* **uniformly dominated** *by T on the elements of E and we use the notation $S \prec\prec T$.*

5.3.1. Examples.

(1) If $S \in Z \circ T$, then clearly $S \prec\prec T$.

(2) Let E and X be vector lattices, and let $S : E \to X$ be a regular d-homomorphism. Then (by Theorem 3.3) there is a lattice homomorphism $T : E \to X$ such that $S \prec\prec T$.

(3) It is worth warning the reader against a possible confusion between the relation " $\prec$ " and the standard order relation " $<$ " for operators. Specifically, let X be a vector lattice and $S, T : X \to X$ two operators such that $0 \leq S \leq T$. It is easy to see that the previous inequalities do not imply in general that $S \prec T$, though of course, $Sx \in \Delta(Tx)$ for each $x \geq 0$.

In the next theorem we deal with the case when X is a vector lattice.

5.4. Theorem. *Let either of the following two conditions be fulfilled.*

(1) *X is an (r_u)-complete vector lattice and $S \prec\prec T$,*

(2) *X is a Banach lattice, $E_1, \ldots, E_n$ are Banach spaces, T is continuous and $S \prec T$.*

Then $S \in Z \circ T$.

Proof. We start with two special cases.

Case I. Let S and T be n-linear functionals, that is, $\dim(X) = 1$. Using induction on n, we will show that $S \prec T$ implies $S \in Z \circ T$. The implication is trivially true if $n = 1$. Let $n > 1$. It is convenient to treat S and T as linear operators defined on the space E_n and with values in the space H of all $(n-1)$-linear functionals defined on $E_1 \times E_2 \times \cdots \times E_{n-1}$. To justify the induction, it is enough to verify the following simple statement:

If for each $e' \in E_n$ there exists a number $\lambda(e')$ such that $Se' = \lambda(e')Te'$, then $S = \lambda_0 T$ for some number λ_0.

To prove this statement, we first note that for each $e' \in E_n$ and for each scalar t, we obviously have $\lambda(te') = \lambda(e')$.

Let e' and e'' be two arbitrary elements from E_n. We want to show that $\lambda(e') = \lambda(e'')$. On one hand, we have $S(e' + e'') = S(e') + S(e'') = \lambda(e')Te' + \lambda(e'')Te''$, and on the other, $S(e' + e'') = \lambda(e' + e'')T(e' + e'') = \lambda(e' + e'')T(e') + \lambda(e' + e'')T(e'')$.

Assuming that Te' and Te'' are linearly independent, we immediately obtain from the previous identities that $\lambda(e') = \lambda(e' + e'') = \lambda(e'')$.

Now let Te' and Te'' be linearly dependent. Without loss of generality we may assume that $Te' = Te'' \neq 0$. Consider now the element $e' - e''$. Since $T(e' - e'') = 0$, we have $S(e' - e'') = 0$. That is, $S(e') = S(e'')$, and hence $\lambda(e')Te' = \lambda(e'')Te''$. This shows that $\lambda(e') = \lambda(e'')$.

Case II. Let X be a Dedekind complete VL. Without loss of generality, we can assume that $(T(E))^{dd} = X$ and that X is embedded in $C_\infty(Q)$, where $Q = Q(X)$ is the Stone space of X. Then, as usual, we can identify $Z(X)$ with $C(Q)$.

First we show that $S \prec T$ implies that there is a function $V \in C_\infty(Q)$ such that $Se = VTe$ for each $e \in E$. To this end, introduce the following set $Q_0 := \{q \in Q : (\exists\, e \in E)\ (0 < (Te)(q) < \infty)\}$. Case I implies that for each $q \in Q_0$ there is a unique number $V_0(q)$ such that $(|(Te)(q)| < \infty) \Rightarrow (Se)(q) = V_0(q)(Te)(q)$. The function V_0 thus defined is obviously continuous on Q_0, and therefore V_0 has a continuous extension $V \in C_\infty(Q)$. Now, for each $e \in E$ and for each q from a dense subset Q_0, we have $(Se)(q) = V(q)(Te)(q)$. This implies that $Se = VTe$.

Assume that condition (1) of the theorem (i.e., that $S \prec\!\prec T$) holds. Then the preceding proof shows immediately that the function V belongs to $C(Q)$ (rather than to $C_\infty(Q)$), and hence $S \in Z \circ T$.

Assume now that we have (2). To show that also in this case $V \in Z$, it is enough to verify that (2) $\Longrightarrow$ (1). It is readily seen that the operator S is continuous with respect to each variable (this follows from the continuity of T, from the closed graph theorem, and from the formula $Se = VTe$, where $e \in E$). For each positive number $a > 0$ introduce the set $E^{(a)} = \{e \in E : |Se| \leq a|Te|\}$. Since S is continuous with respect to each variable, it is jointly continuous (see [Rud], Theorem 2.17) and hence the set $E^{(a)}$ is closed in E. Obviously $E = \cup_{m=1}^{\infty} E^{(a_m)}$ for each sequence $a_m \uparrow \infty$. The Baire category theorem then implies that there is an $a > 0$ such that $E^{(a)}$ has a nonempty interior. We will show that $E^{(a)} = E$, and this will establish condition (1) of the theorem.

If $n = 1$, then it is enough to show that $E^{(a)}$ is a linear subspace of $E = E_1$. Obviously $E^{(a)}$ is closed under scalar multiplication. Let $e_1, e_2 \in E^{(a)}$. The formula $Se = VTe$ $(e \in E)$ shows that $e \in E^{(a)}$ if and only if for each $q \in Q$ the following implication holds $((Te)(q) \neq 0 \Longrightarrow |V(q)| \leq a)$. This clearly implies that $e_1 + e_2 \in E^{(a)}$.

Now let $n > 1$. For each $i = 1, 2, \ldots, n$ there is a nonempty open subset $U_i \subset E_i$ such that $E^{(a)} \supset U_1 \times U_2 \times \ldots \times U_n$. The above result for $n = 1$ implies that $E^{(a)} \supset E_1 \times U_2 \times \ldots \times U_n$, since S and T are linear with respect to the first variable. Applying the same arguments to the second, third, etc variables, we obtain that $E^{(a)} = E$. This completes the proof of the theorem in Case II.

General case. We may (and shall) assume that X is embedded in its Dedekind completion $\hat{X}$. Either of the conditions (1) or (2) of the theorem implies that there exists a $\hat{V} \in Z(\hat{X})$ such that $Se = \hat{V}Te$ for each $e \in E$. Let $Y = \{x \in X : \hat{V}x \in X\}$. If we show that $X_1 = X_{T(E)} \subset Y$ (#), then letting $V := \hat{V}|_{X_1}$ we obviously have $V \in Z(X_1)$ and $S = VT$, that is, the conclusion of the theorem. To verify (#), it will suffice to show that Y is an ideal in X (for, clearly, $T(E) \subset Y$, and therefore Y contains X_1, the ideal generated by $T(E)$). Assume that $|x'| \leq |x|$, where $x' \in X$, and $x \in Y$. We want to show that $x' \in Y$. Since X is (r_u)-complete, there exists an $f \in Z(X_x)$ such that $x' = fx$. Consider f as an operator from X_x to $\hat{X}$. Obviously, it is uniformly dominated by the operator of the canonical embedding $X_x \hookrightarrow \hat{X}$. By Case II there is an $\hat{f} \in Z(\hat{X})$ such that $fy = \hat{f}y$ for $y \in X_x$. Therefore, $\hat{V}x' = \hat{V}fx = \hat{V}\hat{f}x = \hat{f}\hat{V}x = f\hat{V}x \in X$, since $\hat{V}x \in X_x$. This proves that Y is an ideal, and the proof of the theorem is complete. ■

Now we turn our attention to the case of $C(K)$-modules.

5.5. Theorem. *Let X be a $C(K)$-module and let either of the following two conditions be fulfilled.*

(1) *$S \prec\prec T$.*

(2) *$E_1, E_2, \ldots, E_n$ are Banach spaces, $T : E \to X$ is continuous and $S \prec T$.*

Then $S \in Z \circ T$.

Proof. Introduce two $(n+1)$-linear operators $S_1, T_1 : X^* \times E \to L^1_X$ by the following formulas: $S_1(x^*, e) = x^* \square Se$, $T_1(x^*, e) = x^* \square Te$.

Lemma 4.8(3) implies that if the operators S, T satisfy condition (k) of this theorem, where k = 1 or 2, then the operators S_1, T_1 satisfy condition (k) of Theorem 5.4. Therefore, there exists $W \in L^\infty_X = Z(L^1_X)$ such that $S_1 = WT_1$ or (equivalently)

$x^* \square Se = W \cdot x^* \square Te$ for $x^* \in X^*$ and $e \in E$. Let Y denote the submodule generated by $T(E)$ in X. Pick an arbitrary element $\sum_{k=1}^{\ell} f_k Te_k \in Y$, where $f_k \in C(K)$ and $e_k \in E$. Then, for each $x^* \in X^*$, we have $x^* \square \sum_k f_k Se_k = W \cdot x^* \square \sum_k f_k Te_k$. Hence (again by Lemma 4.8(3)) $\sum_k f_k Se_k \in \|W\| \cdot \Delta(\sum_k f_k Te_k)$. Thus we have defined an operator $\sum_k f_k Te_k \mapsto \sum_k f_k Se_k$ on Y, and it satisfies the condition of Corollary 4.10. Therefore it can be extended to an operator $V \in Z(X_Y)$, and this implies the desired formula $S = VT$. ■

6. Some applications. A generalization of Bade's Theorem
Properties of d-homomorphisms

6.1. Definition. *A Banach $C(K)$-module X is said to be an operator module if $X \neq \{0\}$ and the representation of $C(K)$ in $L(X)$ is isometric, i.e., for each function $f \in C(K)$ the norm $\|f\|_{C(K)}$ coincides with the norm of the operator $x \mapsto fx$ in $L(X)$. In this case, we identify $C(K)$ with a closed subalgebra of the algebra $L(X)$.*

6.1.1. For an arbitrary $C(K)$-module X let us introduce the set $J = \{f \in C(K) : fx = 0 \ \forall x \in X\}$, which is evidently an ideal in $C(K)$. Therefore $C(K)/J = C(K_1)$ for an appropriate compact Hausdorff space K_1, and it is easy to verify that X becomes an operator $C(K_1)$-module. (This is a standard fact of the theory of commutative Banach algebras; its generalization to non-commutative C^*-algebras is given in [HO2].) In other words: (i) we can replace a $C(K)$-module X by a naturally associated operator $C(K_1)$-module, and (ii) a $C(K)$-module X is an operator module iff X is exact.

The next theorem is a generalization of a famous theorem of Bade about algebras of spectral operators (see [DS2], Chapter XVII.3). In some sense, our generalization establishes a natural domain of applicability of the Bade Theorem.

6.2. Theorem. (Generalized Bade's Theorem.) *Let X be an operator $C(K)$-module, and let $T : X \to X$ be a linear operator (not a priori assumed to be continuous). The following statements are equivalent.*

(1) *$T \in Z(X)$.*

(2) *T belongs to the closure of $C(K)$ in the strong operator topology.*

(3) *T maps each ideal into itself.*

(4) *T maps each closed ideal (or, equivalently, each closed $C(K)$-invariant subspace) into itself.*

Proof. (1) $\Longrightarrow$ (2). Let us introduce two bilinear operators $S_1, T_1 : X^* \times X \to L^1_X$ by the following formulas: $S_1(x^*, x) = x^* \square Tx$ and $T_1(x^*, x) = x^* \square x$. By Lemma 4.8(3) $S_1 \prec\prec T_1$, i.e., these operators satisfy condition (1) of Theorem 5.4. Hence, there is an operator $W \in Z(L^1_X)$ such that

$$x^* \square Tx = W \cdot x^* \square x, \quad x^* \in X^*, \ x \in X.$$

This implies that whenever $\sum_k x_k^* \square x_k = 0$ for some finite collections $\{x_k^*\} \subset X^*$ and $\{x_k\} \subset X$, then necessarily $\sum_k x_k^* \square T x_k = 0$. In particular this implies that whenever $\sum_k x_k^*(f x_k) = 0$ for all $f \in C(K)$, then $\sum_k x_k^*(T x_k) = 0$.

Recall now (see [DS1], Theorem VI.1.4) that for each linear continuous functional ξ on the space $L(X)$ equipped with the strong operator topology τ_s, there exist some $x_k^* \in X^*$ and $x_k \in X$, $k = 1, 2, \ldots, n$, such that $\xi(S) = \sum_{k=1}^{k=n} x_k^*(S x_k)$ for all $S \in L(X)$. This allows us to reformulate the statement in the previous paragraph as follows: if $\xi(f) = 0$ for each $f \in C(K)$, then $\xi(T) = 0$. Clearly this implies that T belongs to the τ_s-closure of $C(K)$.

(2) $\Longrightarrow$ (4). Since each operator from $C(K)$ maps closed ideals into themselves, the same is true for T, which belongs to the closure of $C(K)$ in the strong operator topology.

(4) $\Longrightarrow$ (3). It is enough to show that for each $x \in X$ the restriction of T to $X(x)$ belongs to the algebra $Z(x)$. It is plain to see that each band in the Banach lattice $X(x)$ is a closed ideal in X; therefore, (4) implies that T maps bands of the vector lattice $X(x)$ into themselves. As shown in [AVK1], the last property implies that $T|X(x) \subset Z(x)$.

(3) $\Longrightarrow$ (1) Let $S_1 = T$ and let T_1 be the identity operator on X. Clearly (3) implies that S_1 and T_1 satisfy condition (2) of Theorem 5.5. Therefore $T \in Z(X)$. ■

6.2.1. Remark. The assumption that X is an *operator* $C(K)$-module is not essential for the validity of the previous theorem. Indeed, if X is an arbitrary $C(K)$-module, then it can be replaced by the associated operator $C(K_1)$-module described in 6.1.1.

6.2.2. Remark. Under the assumption that T is a continuous operator, the equivalence (2) $\Longleftrightarrow$ (4) in Theorem 6.2 was independently obtained by a different method by D. Hadwin and M. Orhon [HO1]. We take this opportunity to thank them for bringing our attention to this and to several other related articles.

6.2.3. Comparison with the Bade theorem. To see that Theorem 6.2 is indeed a generalization of Bade's theorem we begin by recalling a bit of terminology (see for example [HO1] or [Gil]) that will be of use later on, too.

If $\mathcal{S}$ is a subset of $L(X)$, where X is a Banach space, then the symbol Lat$\mathcal{S}$ denotes the collection of all closed subspaces of X which are invariant under each $S \in \mathcal{S}$. The symbol Alg Lat$\mathcal{S}$ denotes the set of all operators in $L(X)$ which leave invariant each subspace in Lat $\mathcal{S}$. Clearly, the set Alg Lat$\mathcal{S}$ is a τ_s-closed subalgebra of $L(X)$ containing both $\mathcal{S}$ and the identity operator. The collection $\mathcal{S}$ is said to be **reflexive** if Alg Lat$\mathcal{S}$ coincides with the unital τ_s-closed subalgebra generated by $\mathcal{S}$.

In this notation Theorem 6.2 says (among other things) that $Z(X) = \tau_s\text{-}cl(C(K))$ $=$ Alg Lat $C(K)$ for each operator $C(K)$-module.

The classical Bade theorem [DS2], Theorem XVII.3.16, asserts that the τ_s-closed algebra generated by a complete Boolean algebra $\mathcal{B}$ of projections in a Banach space X coincides with Alg Lat $\mathcal{B}$. In view of the Stone representation theorem, $\mathcal{B}$ may be identified with the set $\mathcal{S}$ of the characteristic functions of all clopen subsets of an extremally disconnected compact Hausdorff space $Q_{\mathcal{B}}$, the Stone space of $\mathcal{B}$. Obviously the linear span of $\mathcal{S}$ is norm-dense in $C(Q_{\mathcal{B}})$, and consequently X is a $C(Q_{\mathcal{B}})$-module. Therefore, Theorem 6.2 is indeed a broad generalization of Bade's theorem. Moreover, we can show that if X is a $C(K)$-module, then $C(K)$ is generated (in the way described above) by a complete Boolean algebra of projections in X if and only if each ideal $X(x)$, $x \in X$, has order continuous norm. At this point it is also appropriate to mention a paper by A. Veksler [Vek], where it is shown that each Banach space X with a complete Boolean algebra of projections may be turned into a Banach lattice with order continuous norm, provided that there exists a cyclic element in X.

It is well known that the center of an arbitrary Banach lattice is isometrically isomorphic to a $C(S)$-space for an appropriate compact Hausdorff space (see [Wic2] for a short proof of this result). Our next corollary extends this result to arbitrary operator $C(K)$-modules.

6.3. Corollary. *For an arbitrary operator $C(K)$-module X the algebra $Z(X)$ is* *isomorphic to* $C(K_1)$, *where* K_1 *is a compact Hausdorff space. In particular,* $Z(X)$ *is commutative.*

Proof. Being the closure of a $C(K)$-space, the center $Z(X)$ is obviously a commutative algebra. Let us verify that it is a uniform algebra, that is, that $\|T^2\| = 1$ for each $T \in Z(X)$ with $\|T\| = 1$. Fix an arbitrary $\epsilon > 0$ and a normed element $x \in X$ with $\|Tx\| > 1 - \epsilon$. Consider now the restriction of T to the closed ideal $X(x)$. Obviously $T|_{X(x)} \in Z(X(x))$ and the latter is isomorphic to some $C(K_x)$-space, since $X(x)$ is a Banach lattice. Therefore $\|T^2|_{X(x)}\| = \|T|_{X(x)}\|^2 > (1-\epsilon)^2$, whence $\|T^2\| = 1$.

Clearly the natural involution $f \to \bar{f}$ on $C(K)$ satisfies the condition $\|fx\| = \|\bar{f}x\|$, since the elements fx and $\bar{f}x$ of the Banach lattice $X(x)$ have the same modulus. The equivalence (1) $\Longleftrightarrow$ (2) of the previous theorem implies that this involution may be extended to $Z(X)$ and, therefore, it generates in $Z(X)$ a structure of a commutative C*-algebra. But each commutative C*-algebra with a unit coincides with the space of continuous functions on some compact Hausdorff space [Dug],[Tak]. ∎

The next proposition supplements Theorem 6.2 by describing a relationship between the unit balls in $C(K)$ and $Z(X)$. In the case when X is a Hilbert space this result may be inferred from a result due to I. Kaplansky on approximation of self-adjoint elements of the τ_s-closure of self-adjoint algebras of operators (see [Arv], Theorem I.2.2).

6.4. Proposition. *Let* X *be an operator* $C(K)$*-module. Then the unit ball of the algebra* $Z(X)$ *coincides with the closure in the strong operator topology of the unit ball of the algebra* $C(K)$.

Proof. Let $T \in Z(X)$ and $\|T\| \leq 1$. Fix an arbitrary positive integer n and consider the space X^n, the direct sum of n copies of X with the ℓ_1-norm, i.e., $\|\vec{x}\| = \sum_{k=1}^{n} \|x_k\|$, where $\vec{x} = (x_k)$ and $x_k \in X$. Letting $f \cdot (x_1, \cdots, x_n) := (fx_1, \cdots, fx_n)$, $f \in C(K)$, we equip X^n with the structure of a $C(K)$-module. It is plain to see that $L^1_X = L^1_{X^n}$, and that $\vec{x}^* \square \vec{x} = \sum_{k=1}^{n} x_k^* \square x_k$ for arbitrary $\vec{x}^* = (x_1^*, \cdots, x_n^*) \in (X^n)^*$ and $\vec{x} = (x_1, \cdots, x_n) \in X^n$. The operator T generates in a natural way an operator $\tilde{T} : X^n \to X^n$, by the formula $\tilde{T}\vec{x} := (Tx_1, \cdots, Tx_n)$. Introduce also the following two bilinear operators $S_1(\vec{x}^*, \vec{x}) = \vec{x}^* \square \tilde{T}\vec{x}$ and $T_1(\vec{x}^*, \vec{x}) = \vec{x}^* \square \vec{x}$. It is easy to verify that the operators $S_1, T_1 : (X^n)^* \times X^n \to L^1_X$ satisfy the conditions of Theorem 5.4, and hence there is a $W \in L^\infty_X$, $\|W\| \leq 1$ such that $\vec{x}^* \square \tilde{T}\vec{x} = W \cdot (\vec{x}^* \square \vec{x})$ for all $\vec{x}^* \in (X^n)^*$ and $\vec{x} \in X^n$. By Lemma 4.8(3) we have $\tilde{T}\vec{x} \in \Delta(\vec{x})$ for each $\vec{x} \in X^n$. Since n and $\vec{x} = (x_1, \cdots, x_n) \in X^n$ are arbitrary, the conclusion of Proposition 6.4 follows from the definition of the set $\Delta(\vec{x})$. ∎

6.4.1. Below we will present an important corollary to our previous proposition. Let X be an arbitrary Banach lattice and let Z be its center. As we know, $Z = C(K)$ for some compact Hausdorff space K, and consequently X may be considered as a $C(K)$-module. Now, being a $C(K)$-module, X has its own center $Z_{C(K)}(X)$, and the next result shows that (as one might expect) both these centers coincide.

6.4.2. Corollary. $Z_{C(K)}(X) = C(K)$, *i.e., the centers of an arbitrary Banach lattice* X *and of the corresponding* $C(K)$*-module coincide.*

Proof. By Theorem 6.2 the center $Z_{C(K)}(X)$ is the closure of $C(K)$ in the strong operator topology, and therefore, in view of Proposition 6.4, our claim will follow if we show that the unit ball of $C(K)$ is closed in $L(X)$ in the strong operator topology. Since $C(K)$ is the center of the Banach lattice X, for each $T \in C(K)$ the condition $\|T\| \leq 1$ is equivalent to the condition $|Tx| \leq |x|$ for all $x \in X$. Clearly this implies that the unit ball of $C(K)$ is closed in the strong operator topology. ∎

6.5. Now for a while we turn our attention to d-homomorphisms between $C(K)$-modules. Recall that for vector lattices X and Y the symbol $dh(X, Y)$ denotes the set of all regular d-homomorphisms from X into Y, and $dh_+(X, Y) = \{T \in d(X, Y) : T \geq 0\}$ is the set of all positive d-homomorphisms, i.e., lattice homomorphisms. If X is a $C(K_1)$-module and Y is a $C(K_2)$-module, then the same symbol $dh(X, Y)$ will be used to denote the set of all continuous d-homomorphisms from the $C(K_1)$-module X into the $C(K_2)$-module Y.

The following result is an analog for $C(K)$-modules of Theorem 3.3.

6.6. Theorem. *Let* T *be a continuous linear operator from a* $C(K_1)$*-module* X *into a* $C(K_2)$*-module* Y*. The following two conditions are equivalent:*

(1) $T \in dh(X,Y)$;

(2) $x_1 \in \Delta(x_2) \Longrightarrow Tx_1 \in \Delta(Tx_2)$.

Proof. (1) $\Longrightarrow$ (2). Let $x_1 \in \Delta(x_2)$. Fix $y^* \in Y^*$ and consider an operator $U : X(x_2) \longrightarrow L^1_Y$ defined by the following formula: $Ux = y^* \square Tx$, $x \in X(x_2)$. By Lemma 4.8(4), the operator U is a continuous d-homomorphism from $X(x_2)$ into L^1_Y. But each continuous d-homomorphism between Banach lattices is regular. (For real Banach lattices this is proved in [Abr1], and for complex valued lattices it can be reduced to the previous case by decomposing the operator into real and imaginary parts.) Since $x_1 \in \Delta(x_2)$, Theorem 3.3 implies that $Ux_1 \in \Delta(Ux_2)$, or, in other words, $|y^* \square Tx_1| \leq |y^* \square Tx_2|$. A second application of Lemma 4.8(3) allows us to conclude that $Tx_1 \in \Delta(Tx_2)$.

(2) $\Longrightarrow$ (1). Let $x_1, x_2 \in X$ and $x_1 \mathrm{d}\, x_2$. Then, by Lemma 4.8(4), $x_1 \mathrm{d}\, x_2$ in the vector lattice $X(x_1 + x_2)$, and (2) implies that the restriction of T to $X(x_1 + x_2)$ takes values in the vector lattice $Y(T(x_1 + x_2))$ and satisfies the condition (2) of Theorem 3.3. By that theorem, T is a d-homomorphism between the vector lattices $X(x_1 + x_2)$ and $Y(T(x_1 + x_2))$, and hence $Tx_1 \mathrm{d}\, Tx_2$. ■

The next result is a broad generalization, promised in Section 3, of a theorem due to Hart [Har]. This theorem will be generalized in two ways: on the one hand we extend it to $C(K)$-modules, and on the other, in the case of vector lattices, we weaken the assumptions on the spaces in question.

6.7. Theorem. *Let X be a vector lattice and Y be an (r_u)-complete vector lattice, or else let X be a $C(K_1)$-module and Y be a $C(K_2)$-module. Let $T \in dh(X,Y)$ and let Y_1 denote the ideal in Y generated by TX. Then there exists a unique algebraic homomorphism $\gamma : Z(X) \longrightarrow Z(Y_1)$ sending the unit of $Z(X)$ into the unit of $Z(Y_1)$ and such that*

$$T(f \cdot x) = \gamma(f) \cdot Tx \qquad (x \in X,\ f \in Z(X)).$$

Proof. Fix any $f \in Z(X)$ and define an operator $S : X \to Y$ by letting $Sx = Tfx$ for $x \in X$. By Theorem 3.3 or 6.6 (depending on the case in question), there exists a number $c \geq 0$ such that $Tfx \in c\Delta(Tx)$ for all $x \in X$, i.e., $S \prec\prec T$. Therefore, by Theorem 5.4 or 5.5 respectively, we can conclude that $S \in Z \circ T$, and thus, in view of Lemma 5.2, there is a unique element $\gamma(f)$ of the algebra $Z(Y_1)$ such that $S = \gamma(f)T$. Obviously, the mapping $f \longrightarrow \gamma(f)$ is an algebraic homomorphism sending the unit of $Z(X)$ into the unit of $Z(Y_1)$. ■

Remark. In the above cited work [Har], the space Y was assumed to be a Dedekind complete vector lattice. Under this assumption, the algebras $Z(Y_1)$ and $Z((TX)^{dd})$ are clearly canonically isomorphic, and thererfore, Theorem 6.7 implies immediately Hart's theorem.

6.8. As one more illustration of the applicability of the results obtained in Section 5, we are going to present a generalization of a known theorem of S. Kutateladze [Kut] (see also [AB1], Theorem 8.16). We precede it with some discussion.

6.8.1. Recall that if X and Y are vector lattices, then the space of all order bounded (linear) operators from X into Y is denoted by $L_b(X, Y)$.

Let $\hat{Y}$ denote, as usual, the Dedekind completion of Y, and J be the canonical embedding of Y into $\hat{Y}$. Then $L_b(X, \hat{Y})$ is a Dedekind complete vector lattice and for each $T \in L_b(X, Y)$ there exists an operator $|JT| : X \to \hat{Y}$. If $|JT|(X) \subset Y$, then obviously there exists a module $|T| : X \to Y$, and $|JT| = J|T|$. Finally, for each $T \in L_b(X, Y)$, let $\Delta_b(T) = \{S \in L_b(X, Y) : |JS| \leq |JT|\}$. If Y is Dedekind complete, then clearly $\Delta_b(T) = \Delta(T) = \{S \in L_b(X, Y) : |S| \leq |T|\}$.

6.8.2. For the next lemma we need the following result, which follows directly from Theorem 3.3 and which (in the case of real vector lattices) is a reformulation of a theorem due to Meyer [Mey]. *If $T \in dh(X, Y)$, where X, Y are vector lattices, then there exists an operator $|T| : X \to Y$, $|T| \in dh_+(X, Y)$, and $|Tx| = |T|(|x|)$ for each $x \in X$.*

6.8.3. Lemma. *Let $T \in dh(X, Y)$, where X, Y are vector lattices. Then an operator $S \in \Delta_b(T)$ if and only if $(|Sx| \leq |Tx|\ \forall x \in X)$. This implies, in particular, that $S \in dh(X, Y)$.*

Proof. By the definition of the modulus of an operator, the condition $S \in \Delta_b(T)$ is equivalent to the condition $|JSx| \leq |JT|(|x|)$, $x \in X$. By 6.8.2 the last condition is equivalent to the condition $|JSx| \leq |Tx|$, $x \in X$. Thus, if $x_1 \mathrm{d}\, x_2$, then clearly $|JSx_1| \wedge |JSx_2| \leq |Tx_1| \wedge |Tx_2| = 0$, and therefore, $S \in dh(X, Y)$. A second application of 6.8.2 gives the existence of $|S|$ and the equality $|JSx| = |Sx|$. This implies the desired inequality $|Sx| \leq |Tx|$ for all $x \in X$. ∎

6.8.4. Recall that the Kutateladze theorem states that if $T \in dh_+(X, Y)$ and $0 \leq S \leq T$, where X and Y are real vector lattices with Y Dedekind complete, then $S = T_1 T$ for some $T_1 \in Z(Y)$. In our notation it means that

$$T \in dh_+(X, Y) \Longrightarrow \Delta(T) \subset Z \circ T.$$

Now we are in a position to present its generalization, which includes all (r_u)-complete vector lattices and, in particular, all Banach lattices.

6.9. Theorem. *Let $T \in L_b(X, Y)$, where X and Y are vector lattices. The following statements are true.*

(1) *If Y is (r_u)-complete and $T \in dh(X, Y)$, then $\Delta_b(T) \subset Z \circ T$.*

(2) *If X is (r_u)-complete and $\Delta_b(T|X_x) \subset Z \circ (T|X_x)$ for each $x \in X$, then the operator $T \in dh(X, Y)$.*

(3) *If X is Dedekind complete and $\Delta_b(T) \subset Z \circ T$, then $T \in dh(X, Y)$.*

Proof. The first statement follows from Lemma 6.8.3 and Theorem 5.4. Now let the conditions of Statement (2) be satisfied. Pick up $x_1, x_2 \in X$ with $|x_1| \leq |x_2|$. Since X is (r_u)-complete, $x_1 = f x_2$ for some $f \in Z(X_{x_2})$ with $|f| \leq 1$. It is plain to see that the operator $x' \mapsto T(fx')$ defined on X_{x_2} belongs to $\Delta_b(T|X_{x_2})$, and therefore there exists $g = \gamma(f) \in Z(Y_{T(X_{x_2})})$ such that $Tfx' = gTx'$ for each $x' \in X_{x_2}$. The correspondence $f \mapsto \gamma(f)$ clearly generates an algebraic homomorphism, and thus the condition $|f| \leq 1$ implies $|g| = |\gamma(f)| \leq 1$. This shows that $|x_1| \leq |x_2|$ implies that $|Tx_1| \leq |Tx_2|$. By Theorem 3.3 we can conclude that $T \in dh(X, Y)$,that is, the conclusion of Statement (2) is proved.

Finally, let the conditions of Statement (3) hold. Take $x \in X$ and $S \in \Delta_b(T|X_x)$. In view of (2), to prove (3) it is enough to prove that $S \in Z \circ (T|X_x)$. Let P denote the band projection from X onto the band $\{x\}^{dd}$. Clearly, the operator $x' \mapsto S(Px')$, $x' \in X$, belongs to $\Delta_b(T)$, and therefore it belongs to $Z \circ T$. This immediately implies that $S \in Z \circ (T|X_x)$. ■

7. The center of the conjugate space

In this section, we continue the investigation of the parallelism between $C(K)$-modules and Banach lattices.

Let X be a Banach lattice. Then X^* is likewise a Banach lattice with its natural mapping $x^* \mapsto \Delta(x^*)$. Recall that X^* is always Dedekind complete. Let $K_1 = Q(X^*)$ be its Stone space. Then the center $Z(X^*)$ of the Banach lattice X^* is (isomorphic to) $C(K_1)$, and therefore X^* becomes an operator $C(K_1)$-module. The structure of this $C(K_1)$-module generates its own mapping $x^* \mapsto \Delta(x^*)$, which obviously coincides with the initial one. As it follows from Corollary 6.4.2 the center of the $C(K_1)$*-module* X^* coincides with $C(K_1)$, that is, with the center of the *Banach lattice* X^*. In short, $Z_{C(K_1)}(X^*) = C(K_1)(= Z(X^*))$.

7.1. In this section, a mapping $x^* \mapsto \Delta(x^*)$ and an algebra $Z(X^*)$ will be defined for an **arbitrary** $C(K)$-module X. Also, for an arbitrary Banach lattice X, we will define a bilinear mapping $(x^*, x) \mapsto x^* \,\square\, x$, and the spaces L^1_X and L^∞_X; and, as we will show, the order and algebraic properties of these objects (for Banach lattices and $C(K)$-modules) are absolutely similar.

7.2. It is plain to see that if X is an operator $C(K)$-module, then X^* is likewise a $C(K)$-module, provided one defines the action on X^* in a natural way as

$$f: \; x^* \mapsto f^* x^*,$$

where f^* is the conjugate of $f \in C(K) \subset L(X)$. Consequently one can talk about the center of this module. Therefore, the reader may be tempted to ask why we then need a special definition for the center of the conjugate space. The point is that the above suggested definition, though natural and sometimes useful (see for instance 9.14.3 and Proposition 9.8) does not allow us to preserve the desired parallelism between the theory of $C(K)$-modules and that of Banach lattices.

We begin with a formal definition of a new mapping (temporarily denoted by) $x^* \mapsto \blacktriangle(x^*)$ on X^* using only the fact that X^* is the conjugate of a space X carrying the mapping $x \mapsto \Delta(x)$. As soon as we show in Theorem 7.6 that in the case of a $C(K)$-

module X this new mapping on X^* coincides with the interval mapping $x^* \mapsto \Delta(x^*)$ on the $C(K_1)$-module X^* defined there, we will use only the latter notation.

7.3. Definition. *Let X be an arbitrary Banach lattice or a $C(K)$-module and $x^* \in X^*$. Then let*

$$\blacktriangle(x^*) := \{y^* \in X^* : |y^*(x)| \leq \sup_{y \in \Delta(x)} |x^*(y)| \;\; \forall x \in X\}.$$

To justify the usage of the triangular symbol $\blacktriangle(x^*)$, we start by verifying that for any Banach lattice X the just introduced mapping $x^* \mapsto \blacktriangle(x^*)$ coincides with the canonical one $x^* \mapsto \Delta(x^*)$ existing on the Banach lattice X^*.

7.4. Theorem. *Let X be an arbitrary Banach lattice. Then $\blacktriangle(x^*) = \Delta(x^*)$ for each $x^* \in X^*$.*

Proof. Take $y^* \in X^*$. The inclusion $y^* \in \Delta(x^*)$, that is, the inequality $|y^*| \leq |x^*|$ is equivalent to the condition that for each $x \in X$ we have $|y^*(x)| \leq |x^*|(|x|) = \sup_{y \in \Delta(x)} |x^*(y)|$, and this is precisely the statement that $y^* \in \blacktriangle(x^*)$. ■

7.5. Definition. (1) *If X is a Banach lattice, then denote by $Q(X^*)$ the Stone space of the Dedekind complete Banach lattice X^*.*

(2) *If X is a $C(K)$-module, then denote by $Q(X^*)$ the Stone space of the Dedekind complete space L_X^∞ or, equivalently, the maximal ideal space of the algebra L_X^∞.*

7.6. Theorem. *Let X be a Banach lattice or a $C(K)$-module, and let $K_1 := Q(X^*)$. Then X^* can be turned into an operator $C(K_1)$-module in such a way that for each functional $x^* \in X^*$ the set $\blacktriangle(x^*)$ defined in Definition 7.3 coincides with the set $\{hx^* : h \in C(K_1), \|h\| \leq 1\}$, that is, $\blacktriangle(x^*) = \Delta_{C(K_1)}(x^*)$.*

Proof. If X is a Banach lattice, then the conclusion of the theorem is obvious in view of Theorem 7.3 and in view of the above mentioned canonical isomorphism $Z(X^*) = C(K_1)$.

Now let X be a $C(K)$-module. By Definition 7.5(2), $C(K_1)$ may be identified with L^∞_X. Since L^1_X is an AL-space and $L^\infty_X = Z(L^1_X)$, the spaces L^∞_X and L^1_X are in such a duality $<\cdot,\cdot>$ that $<h_1h_2,\mu>=<h_1,h_2\mu>$ for all $h_1,h_2 \in L^\infty_X$, $\mu \in L^1_X$.

For an arbitrary $h \in L^\infty_X$ and arbitrary $x^* \in X^*$ we denote by $h \cdot x^*$ (or simply hx^*) the following functional on X, i.e., an element from X^* :

$$(hx^*)(x) \ = \ < h,\ x^* \,\square\, x > . \tag{$\star$}$$

We claim that definition $(\star)$ turns X^* into a L^∞_X-module. To prove it we need to verify that $\mathbf{1} \cdot x^* = x^*$ and that $h_1(h_2x^*) = (h_1h_2)x^*$. The first property is obvious, and a straightforward verification of the second depends on the following identity (which will also be needed later on):

$$h \cdot (x^* \,\square\, x) = (hx^*)\square\, x, \tag{$\star\star$}$$

where $h \cdot (x^* \,\square\, x)$ is the usual product in L^1_X of the measure $x^* \,\square\, x$ and the function $h \in L^\infty_X = (L^1_X)^*$, and $hx^* \in X^*$ is defined by $(\star)$. (It is worth noticing that $(\star\star)$ is an analogue of identity (4.1) in 4.13.1.)

To verify $(\star\star)$ it is enough to show that for each $f \in C(K)$ we have the equality

$$< h \cdot (x^* \square x), f > \ = \ < (hx^*)\square x, f > .$$

Consider the left term:

$$< h \cdot (x^* \,\square\, x), f > = \int f \, d(h \cdot (x^* \,\square\, x)) = \int fh \, d(x^* \,\square\, x) =$$

$$\int h \, d(f \cdot (x^* \,\square\, x)) = \text{by } (4.1) \ = \int h \, d(x^* \,\square\, fx) = < h, x^* \,\square\, fx > .$$

Now evaluate the right term:

$$< (hx^*)\square x, f > = \text{by def. of } \square \ = (hx^*)(fx) = \text{by } \ (\star) \ = < h, x^* \square\, fx > .$$

This proves $(\star\star)$.

A routine verification shows that the norm of the operator $x^* \mapsto hx^*$ is less than or equal to $\|h\|_{L^\infty_X}$. This establishes that X^* is an L^∞_X-module. To verify that X^*

is an operator module it will suffice to show (see 6.1.1) that this module is exact, i.e., if $0 \neq h \in L_X^\infty$, then $x^* \mapsto hx^*$ is a non-zero operator on X^*. If not, then $hx^*(x) = < h, x^* \square x >= 0$ for all $x^* \in X^*$, $x \in X$, and thus $< h, \mu >= 0$ for all $\mu \in L_X^1$. Hence $h = 0$, a contradiction.

Now, if $y^* = hx^*$ and $\|h\|_{L_X^\infty} \leq 1$, then $y^* \in \Delta_{C(K_1)}(x^*)$, and in view of $(\star\star)$ we have $|y^* \square x| \leq |x^* \square x|$ for all $x \in X$. This implies that $y^* \in \blacktriangle(x^*)$. In other words we have shown that $\Delta_{C(K_1)}(x^*) \subset \blacktriangle(x^*)$.

Conversely, let $y^* \in \blacktriangle(x^*)$. Substituting fx for x in the inequality $|y^*(x)| \leq \sup_{y \in \Delta(x)} |x^*(y)|$, where $f \in C(K)$, $\|f\| \leq 1$, and taking the supremum over all such f we obtain the following inequality: $\|y^* \square x\| \leq \|x^* \square x\|$. Again substituting fx for x and using the fact that f is arbitrary, we obtain $|y^* \square x| \leq |x^* \square x|$. The last inequality implies that there exists an $h \in L_X^\infty$, with $\|h\| \leq 1$ for which $y^* \square x = h \cdot (x^* \square x)$. By $(\star\star)$ we have $h \cdot (x^* \square x) = (hx^*) \square x$, and thus, since x is arbitrary, we can conclude that $y^* = hx^*$, i.e., $y^* \in \Delta_{C(K_1)}(x^*)$. This completes the proof of the theorem. ■

7.6.1. Remark. Let X be a $C(K)$-module and fix an arbitrary $f \in C(K)$ and $x^* \square x \in L_X^1$. As we know (see (4.1) in 4.13.1)

$$f \cdot (x^* \square x) = x^* \square fx.$$

On the other hand (since the space $C(K)$ may be considered as being embedded in $L_X^\infty = C(K_1)$), the element f belongs also to L_X^∞, and thus by $(\star\star)$

$$f \cdot (x^* \square x) = (fx^*) \square x.$$

This possibility of two different interpretations does not cause any ambiguity since these two measures coincide. To prove it we pick an arbitrary $g \in C(K)$ and show that both measures take the same value at the function g. We have

$$g(x^* \square fx) \overset{\square}{=} x^*(gfx)$$

and

$$g(fx^* \square x) \overset{\square}{=} (fx^*)(gx) \overset{(\star)}{=} < f, x^* \square gx > \overset{\square}{=} x^*(fgx).$$

Since $gfx = fgx$ and g is arbitrary, we are done. (By placing a square above the equality sign $\overset{\square}{=}$ we point out that Definition 4.7 of the operation $\square$ is used here.) ■

It is shown in Theorem 7.6 that the space X^* is an operator $C(K_1)$-module. The next result asserts that the center $Z_{C(K_1)}(X^*)$ of this module coincides with $C(K_1)$ itself. This establishes that for the conjugate of an arbitrary $C(K)$-module X the situation is an exact replica of that for the conjugate of a Banach lattice X (see the discussion at the very beginning of this section).

7.7. Proposition. *Under the above notation,* $Z_{C(K_1)}(X^*) = C(K_1)$.

Proof. We have a natural one-to-one homomorphism $\eta : C(K_1) \to Z_{C(K_1)}(X^*)$ defined in the course of the proof of Theorem 7.6 by $\eta(h)(x^*) = hx^*$. We claim that η is surjective. Fix an arbitrary operator $H \in Z_{C(K_1)}(X^*)$ with $\|H\| \leq 1$. By Proposition 6.4 there exists a net $h_\alpha \in C(K_1) = L_X^\infty = (L_X^1)^*$ such that $\|h_\alpha\| \leq 1$ and $\|h_\alpha x^* - Hx^*\|_{X^*} \to 0$ for each $x^* \in X^*$. We will show that there is an $h \in C(K_1)$ such that $\eta(h) = H$.

First of all, let us prove that $\{h_\alpha\}$ is $\sigma(L_X^\infty, L_X^1)$-Cauchy. Since the norms of h_α are uniformly bounded and linear combinations of elementary measures $x^*\square\, x$ are dense in L_X^1, it is enough to show that the net of scalars $h_\alpha(x^*\square\, x)$ converges for each $x^*\square\, x$.

We have

$$h_\alpha(x^*\square\, x) \overset{(*)}{=} (h_\alpha x^*)(x) \to (Hx^*)(x). \qquad (*)$$

Hence indeed, $\{h_\alpha\}$ is a Cauchy net. Denote by h its $\sigma(L_X^\infty, L_X^1)$-limit. It follows from $(*)$ that this h satisfies the following relations

$$h(x^*\square\, x) = (hx^*)(x) = (Hx^*)(x)$$

for all $x^* \in X^*$ and $x \in X$. This implies immediately that $hx^* = Hx^*$ for each $x^* \in X^*$. But this means that $\eta(h) = H$. ■

7.7.1. Corollary. *Let X be a $C(K)$-module. Then the closed unit ball of the algebra $Z(X^*)$ coincides with the closure of the unit ball U of $C(K)$ in the topology $\sigma(L(X^*), X^* \otimes X)$.*

Proof. Our first claim is that for each $h \in Z(X^*)$ with $\|h\| \leq 1$ and each $x^* \in X^*$ there exists a net $\{f_\alpha\} \in U$ such that $f^*x^* \to hx^*$ in the topology $\sigma(X^*, X)$. If not, then by a standard separation theorem we can find an element $x \in X$ such that $\sup_{f \in U} |x^*(fx)| < |hx^*(x)|$. This contradicts the previous proposition in view of Definition 7.3 and Theorem 7.6.

Now, if we introduce a new $C(K)$-module, which is the ℓ_1-direct sum of n-copies of X, and apply to this module the first claim, we get the desired conclusion. ■

7.8. Definition. *Let X be a Banach lattice or a $C(K)$-module, $K_1 = Q(X^*)$. From now on we always consider X^* as a $C(K_1)$-module, $Z(X^*)$ denotes the center of this $C(K_1)$-module X^* and, as we have shown, $Z(X^*)$ coincides with $C(K_1)$.*

If X is a Banach lattice, then this definition of $Z(X^*)$ is equivalent to the definition discussed at the beginning of the section. If X is a $C(K)$-module, then we may say that the algebra L_X^∞, which is isomorphic to $C(K_1)$, acts on X^*, and thus L_X^∞ may be identified with $Z(X^*)$. Therefore we can consider L_X^1 as a subspace of the space $Z(X^*)^* = (L_X^\infty)^*$, and consequently by $(\star)$, we get $(x^* \square x)(h) = (hx^*)(x)$ for each $h \in Z(X^*)$, $x^* \in X^*$, and $x \in X$.

7.9. In our pursuit for a parallelism between Banach lattices and $C(K)$-modules our next step gives a feedback of the latter theory to the former one. Specifically, for an *arbitrary Banach lattice* X we are now able to define a bilinear mapping $x^* \square x$ on $X^* \times X$, and with its help to define the "right" objects L_X^1 and L_X^∞.

7.10. Definition. *Let X be an arbitrary Banach lattice and $x^* \in X^*$, $x \in X$. We denote by $x^* \square x$ the following linear continuous functional on $Z(X^*)$*

$$(x^* \square x)(h) := (hx^*)(x), \quad h \in Z(X^*).$$

7.10.1. Further, we denote by L_X^1 the band in $Z(X^*)^*$ generated by all functionals $x^* \square x$, where $x^* \in X^*$ and $x \in X$. Since $Z(X^*) = C(K_1)$, the space L_X^1 is an AL-space. Its center $Z(L_X^1)$ will be denoted by L_X^∞. The elements of the space L_X^1 may be interpreted as measures on the compact space $K_1 = Q(X^*)$.

Imitating the proof of Lemma 4.11.1 it is easy to verify that L_X^1 coincides with the norm closure in $Z(X^*)^*$ of the linear subspace generated by all functionals $x^* \square x$.

7.10.2. Lemma. *$Z(L^1_X)$ may be identified with $C(K_1)$.*

7.10.3. Remark. If X is a Banach lattice, then the spaces L^1_X and L^∞_X introduced above admit the following convenient description. It is well known (see, for example, [KA]) that there exists a "good" measure space (Ω, Σ, μ) (satisfying the direct sum property), and order isomorphic embeddings of X^* and X into the space $S = S(\Omega, \Sigma, \mu)$ of all (equivalence classes of) measurable functions on Ω, such that X^* is an order dense ideal in S, X is an order dense sublattice of S, and the duality between X^* and X is given by the integral:

$$x^*(x) = \int x^* \cdot x d\mu,$$

where $x^* \cdot x$ denotes the usual multiplication of functions in S. It is readily seen that in terms of this representation we have $Z(X^*) = L^\infty(\Omega, \Sigma, \mu)$,
$L^1_X = L^1(\Omega, \Sigma, \mu)$, and $x^* \square x = x^* \cdot x$. We will refer to this representation as the *canonical representation of the pair* (X^*, X) on (Ω, Σ, μ).

The above remark shows, in particular, that for a Banach lattice X the algebras $Z(X^*)$ and L^∞_X may indeed be identified, thus giving a proof of Lemma 7.10.2. (Recall that for a $C(K)$-module X the identification between $Z(X^*)$ and L^∞_X is a direct consequence of Definitions 7.8 and 7.5).

Further we will need the following analog of Lemma 4.8.

7.11. Lemma. *Let X be any Banach lattice and let x^*, $y^* \in X^*$. The following statements are true.*

(1) *An element $x \in X$ (as a functional on X^*) is orthogonal to $X^*(x^*)$ if and only if $x^* \square x = 0$.*

(2) *An element $x \in X$ (as a functional on X^*) is positive on $X^*(x^*)_+$ if and only if $x^* \square x \geq 0$.*

(3) $y^* \in \Delta(x^*)$ $\Leftrightarrow$ $\|y^* \square x\| \leq \|x^* \square x\|$ *for all* $x \in X$ $\Leftrightarrow$ $|y^* \square x| \leq |x^* \square x|$ *for all* $x \in X$.

(4) $x^* \mathrm{d}\, y^*$ $\Leftrightarrow$ *the elements x^*, y^* belong to the lattice $X^*_{x^*+y^*}$ and are disjoint in it* $\Leftrightarrow$ $(x^* \square x)\ \mathrm{d}\ (y^* \square x)$ *for all* $x \in X$.

All these statements either are easily verifiable or have been already checked in the process of the proof of Theorem 7.6. ■

Making use of the isomorphism between $Z(X^*)$ and L_X^∞ and applying the last statement of the previous lemma, we can easily prove the following theorem.

7.12. Theorem. *Let X be a Banach lattice or a $C(K)$-module. Then the Boolean algebra $\mathcal{B}(X^*)$ of all band projections in $Z(X^*)$ is (order) complete. For arbitrary $x^* \in X^*$, $h \in L_X^\infty$ and $x \in X$ there exist projections P_{x^*}, P_h and P_x, the smallest projections among the projections P* in *$\mathcal{B}(X^*)$ which (respectively) satisfy the following conditions: $P_{x^*}x^* = x^*$, $P_h h = h$ and $P_x(y^* \square x) = y^* \square x$ for each $y^* \in X^*$. Two elements x^*, $y^* \in X^*$ are disjoint if and only if $P_{x^*}P_{y^*} = 0$. Two elements x, $y \in X$ are disjoint if and only if $P_x P_y = 0$.*

8. An analog of the factorization theorem of G. Lozanovsky

Let X be a Banach function space on a measure space (Ω, Σ, μ) and let X' be its dual space, that is,

$$X' = \{x' \in S(\Omega, \Sigma, \mu) : x'x \in L^1(\Omega, \Sigma, \mu) \ \ \forall x \in X\}.$$

The reader is referred to [KA] for relevant information about dual spaces. G. Ya. Lozanovsky proved the following remarkable result [Loz], Theorem 6. For an arbitrary function $h \in L^1(\Omega, \Sigma, \mu)$ and for an arbitrary positive number $\epsilon > 0$ there exist two elements $x' \in X'$ and $x \in X$, such that

$$h = x' \cdot x \text{ and } \quad \|x'\|_{X'} \cdot \|x\|_X \leq (1 + \epsilon)\|h\|_{L^1}.$$

This result has numerous applications and it has since been reproved by many authors. We are going to present an analog of this theorem for an arbitrary Banach lattice or a C(K)-module X, replacing $L^1(\Omega, \Sigma, \mu)$ by L^1_X, and accordingly replacing the dual X' by the Banach conjugate X^*. Important applications of this theorem will be given in the next section.

8.1. Theorem. (Factorization Theorem) *Let X be an arbitrary Banach lattice or a $C(K)$-module and let $m \in L^1_X$. Then for each $\epsilon > 0$, there are $x^* \in X^*$ and $x \in X$ such that $m = x^* \square x$ and $\|x^*\| \cdot \|x\| \leq (1 + \epsilon)\|m\|$.*

The proof of this theorem will be divided into several lemmas.

8.2. Lemma. *If X is a Banach lattice, then the conclusion of Theorem* 8.1 *is true.*

Proof. Let us assume that the pair (X^*, X) is represented on (Ω, Σ, μ) in the sense described in Remark 7.10.3. Denote by W the ideal in $S(\Omega, \Sigma, \mu)$ generated by X and equip it with the following lattice norm:

$$\|w\| = \inf\{\|x\| : x \in X_+, \ |w| \leq x\}.$$

The only axiom for the lattice norm which we need to verify is that $\|w\| = 0$ implies $w = 0$. Other axioms are obvious. Pick $x_n \in X_+$ such that $\|x_n\| < 1/n$ and $|w| \leq x_n$.

and $|w| \le x_n$. It is well-known that the norm convergence in any Banach lattice is equivalent to $(*)(r_u)$-convergence (see for example [Vul], Theorem 7.2.1), and therefore we can find a subsequence $\{x_{n_k}\}$ of $\{x_n\}$ which (r_u)-converges to zero. This means that there exists an element $r \in X$ such that for each $\varepsilon > 0$ the inequality $x_{n_k} \le \varepsilon r$ holds provided k is large enough. Certainly this implies that $w = 0$.

Thus W is a Banach function space on (Ω, Σ, μ). Let us verify that its dual W' coincides with X^*. First notice that the restriction operator $w' \to w'|_X$ is a linear isometry of W' into X^*. On the other hand, if $x^* \in X^*$, we can define $w' \in W'$ by $w'(w) = \int x^* w \, d\mu$, and then clearly $w'|_X = x^*$, i.e., $W' = X^*$.

Fix an $\epsilon > 0$. By Lozanovsky's theorem cited above, there are $y^* \in X^* = W'$ and $w \in W$ such that $m = y^* \cdot w$ and $\|y^*\|_{X^*} \|w\|_W < (1+\epsilon)\|m\|$. By the definition of W, there exists an $x \in X_+$ such that $\|y^*\|_{X^*} \|x\|_W < (1+\epsilon)\|m\|$ and $|w| \le x$. The last inequality implies that $w = hx$ for some function $h \in L^\infty(\Omega, \Sigma, \mu)$. Letting $x^* := h \cdot y^*$, we obtain both the representation $m = x^* \cdot x = x^* \square x$ and the desired estimate of the norm of $\|m\|$. ■

8.2.1. Remark. In the process of the proof above we have used a special case of the following useful observation.

Let X be a vector lattice, $X^\sim$ its order dual, $\pi : X \to (X^\sim)^\sim_n$ the canonical embedding, and W the ideal in $(X^\sim)^\sim_n$ generated by $\pi(X)$. Then $W^\sim_n = X^\sim$.

8.3. Lemma. *Let X be a $C(K)$-module and $\mu = x^* \square x$ be an element from $C(K)^*$ for some $x^* \in X^*$, $x \in X$. Then for each measure m, which is absolutely continuous with respect to μ, the conclusion of Theorem 8.1 is true.*

Proof. Since L^1_X is a band in $C(K)^*$, necessarily $m \in L^1_X$. Let y^* denote the restriction of the functional x^* to the ideal $Y = X(x)$. We can consider Y as a Banach lattice and clearly $\mu = y^* \square x$. Thus, $m \in L^1_Y$. By the previous lemma $m = y_1^* \square y$ for some $y_1^* \in Y^*$, $y \in Y$ satisfying the desirable estimate for the norm. Let x_1^* be any norm preserving extension of y_1^* to X. It can be readily seen that $m = x_1^* \square y$. ■

8.4. Lemma. *Let X be a $C(K)$-module and $m \in L^1_X$. Then we can find a functional $x^* \in X^*$ and a sequence $\{x_k\}$ of elements in X, such that $\sum_k \|x_k\| < \infty$,* *and the measure m is absolutely continuous with respect to the measure* $\sum_k |\mu_k|$, *where* $\mu_k = x^* \mathbin{\square} x_k$.

Proof. By Corollary 4.12.4 there exist two sequences $\{x_k^*\} \subset X^*$ and $\{x_k\} \subset X$ such that $\sum_k \|x_k^*\| < \infty$, $\sum_k \|x_k\| < \infty$, the measures $x_k^* \mathbin{\square} x_k$ are mutually disjoint, and the measure m is absolutely continuous with respect to the measure $\sum_k x_k^* \mathbin{\square} x_k$. Let P_k $(k = 1, 2, \cdots)$ denote the band projection in L_X^1 on the band generated by $x_k^* \mathbin{\square} x_k$. Then $P_k \in L_X^\infty$, and thus by Proposition 7.7 we can consider P_k as an operator on X^*. Letting $x^* = \sum_k P_k x_k^*$ and noticing that for each k, the measure $x_k^* \mathbin{\square} x_k$ is absolutely continuous with respect to the measure $x^* \mathbin{\square} x_k = \sum_j P_j(x_j^* \mathbin{\square} x_k)$, we come to the desired conclusion. ■

The next lemma is basically known (cf. [AS], Proposition 2.4(2), and [Dre], Theorem 10.7).† For the completeness we are including an independent proof.

8.5. Lemma. *Let $\{\mu_k\}$ be a sequence of complex valued measures on a measure space Ω and let $\mu = \sum_k |\mu_k|$ be finite. Then there is a bounded sequence $\{\epsilon_k\}$ of positive numbers such that the measure μ is absolutely continuous with respect to the measure $\sum_k \epsilon_k \mu_k$.*

Proof. By the Radon-Nikodym theorem, $\mu_k = f_k \mu$ for some $f_k \in L^1(\mu)$. For each $n \in \mathbf{N}$, introduce the set $F_n = \{\omega \in \Omega : \sum_{k=1}^{n} |f_k(\omega)| > 0\}$ and let $h_n := \chi_{F_n} \sum_{k=n+1}^{\infty} |f_k|$. The sequence $\{\epsilon_k\}$ will be defined inductively and in such a way that for each n, the function $g_n := |\sum_{k=1}^{n} \epsilon_k f_k|$ is positive μ-almost everywhere on F_n $(*)$.

Set $\epsilon_1 = 1$ and assume that some positive numbers $\epsilon_1, \cdots, \epsilon_n$ have been already chosen to satisfy $(*)$. Introduce further a set A_n, consisting of all those numbers ϵ for which the function $g_{n+1} = |g_n + \epsilon f_{n+1}|$ equals zero on a proper subset $F_n(\epsilon)$ of F_n having a positive measure. Obviously $\mu(F_n(\epsilon') \cap F_n(\epsilon'')) = 0$ provided $\epsilon' \neq \epsilon''$. Consequently A_n is at most countable, and therefore we can find an ϵ_{n+1} such that (1)

† We take this opportunity to thank Z. Lipecki for providing us with the appropriate references.

$0 < \epsilon_{n+1} \leq \epsilon_n$, (2) $\epsilon_{n+1} \notin A_n$, and (3) $\mu\{\omega \in \Omega : g_n(\omega) < \epsilon_{n+1} h_n(\omega)\} \leq 2\mu(\Omega \backslash F_n)$. (To satisfy (3), it is enough to take ϵ_{n+1} small enough.)

It is easy to verify now that this choice of numbers ϵ_k implies that the function $g := |\sum_{k=1}^{\infty} \epsilon_k f_k|$ is positive μ-almost everywhere on Ω. Indeed, for each n, we have

$$g \geq \chi_{F_n} g \geq g_n - \chi_{F_n} \sum_{k=n+1}^{\infty} \epsilon_k |f_k|$$
$$\geq g_n - \epsilon_{n+1} \chi_{F_n} \sum_{k=n+1}^{\infty} |f_k| = g_n - \epsilon_{n+1} h_n.$$

Therefore (3) implies that $\mu\{\omega \in \Omega : g(\omega) \leq 0\} \leq 2\mu(\Omega \backslash F_n) \longrightarrow 0$ as $n \to \infty$. ■

8.6. Lemma. *Let X be a $C(K)$-module and $m \in L^1_X$. Then the conclusion of Theorem* 8.1 *is true.*

Proof. Pick up a sequence $\{\mu_k\}$ of measures $\mu_k = x_k^* \square x_k$ in accordance with Lemma 8.4 and after that a sequence $\{\epsilon_k\}$ in accordance with Lemma 8.5. Thus the measure m is absolutely continuous with respect to the measure $x^* \square \sum_k \epsilon_k x_k$. An application of Lemma 8.3 completes the proof. ■

8.7. Remark. Recall that a norm on a Banach lattice is said to be semi-continuous if $\lim \|x_\alpha\| = \|x\|$, whenever $0 \leq x_\alpha \uparrow x$. In the same paper [Loz], G. Lozanovsky proved also that if the norm on the Banach function space X is semi-continuous, then for every $h \in L_1(\Omega, \Sigma, \mu)$ there is a decomposition $h = x' \cdot x$ such that $\|x'\|_{X'} \cdot \|x\|_X = \|h\|_{L^1}$. It is possible to show that a corresponding version of this result holds also for $C(K)$-modules with semi-continuous norm, which are introduced in Definition 12.10.1.

9. Special Algebras of Operators on X^*
A dual Bade Theorem

Let X be a $C(K)$-module or a Banach lattice. Recall that according to Definition 7.8, in either case X^* is endowed with the structure of $C(K_1)$-module, where $K_1 = Q(X^*)$. In this section, we will be primarily interested in the investigations of the relations between the following subalgebras of the algebra $L(X^*)$ of all linear continuous operators on the conjugate space X^*.

$Z^*(X) = \{T^* : T \in Z(X)\}$†;

$Z(X^*)$, the center of the conjugate space X^*;

$W(X^*) = \{S \in L(X^*) : SE \subseteq E$ for each w^*-closed ideal E of $X^*\}$;

$Z(X^*)^{com}$, the commutant of $Z(X^*)$ in $L(X^*)$;

$(Z^*(X))^{com}$, the commutant of $Z^*(X)$ in $L(X^*)$.

Some of these algebras are just the "conjugate counterparts" of the familiar subalgebras of $L(X)$, and some, generated by a richer structure of the conjugate space, are new.

9.1. Several relations between these algebras are immediate, and we are going to start with them. After that we will be in a position to pose and discuss the main problems considered in this section.

9.1.1. Here we will show that $Z^*(X) \subset Z(X^*)$. To show this, let us first verify that for each $T \in Z(X)$ and for all $x^* \in X^*$, $x \in X$ we have ‡

$$(T^*x^*) \square x = x^* \square Tx. \tag{9.1}$$

Indeed for an arbitrary $f \in C(K)$ we have

$$< T^*x^* \square x, f > \overset{\square}{=} < T^*x^*, fx > = < x^*, T(fx) > \overset{\text{Cor.6.3}}{=}$$
$$= < x^*, f(Tx) > \overset{\square}{=} < x^* \square Tx, f >,$$

† This algebra $Z^*(X)$ of the conjugate operators should not be confused with the conjugate space $Z(X)^*$ of the center $Z(X)$. We hope this similarity in notations will not cause any ambiguity.

‡ This is a generalization of the identity $(f^*x^*) \square x = x^* \square fx$ proved at the end of 4.13.1 for each $f \in C(K)$.

which proves (9.1).

To finish the proof of the inclusion, note that if $T \in Z(X)$, then for an arbitrary $x \in X$, we have $Tx \in \lambda\Delta(x)$ where $\lambda = \|T\|$, and thus, by Lemma 4.8(3), the inequality $\|x^* \square Tx\| \leq \lambda\|x^* \square x\|$ holds for all $x^* \in X^*$. In view of (9.1) this can be rewritten as $\|(T^*x^*) \square x\| \leq \lambda\|x^* \square x\|$, and hence, by Lemma 7.11(3) we have $T^*x^* \in \lambda\Delta(x^*)$. That is, $T^* \in Z(X^*)$. ■

9.1.2. The inclusion $Z^*(X) \subset Z(X^*)$ is proper, in general. For example, this is true for the Banach lattice $X = C[0,1]$.

9.1.3. Another immediate consequence of 9.1.1 is the inclusion $(Z(X^*))^{com} \subset (Z^*(X))^{com}$.

9.1.4. Since each operator T from the center $Z(X^*)$ sends every ideal to itself, we have that $Z(X^*) \subset W(X^*)$.

9.1.5. The inclusion in 9.1.1 gives rise to our first problem.

Problem 1. Under what conditions on X do we have the equality $Z^*(X) = Z(X^*)$?

This condition is found in Theorem 9.10 and this condition reads that for each $x \in X$ the norm in the Banach lattice $X(x)$ must be order continuous.

The observation in 9.1.3 motivates our next problem.

Problem 2. Under what conditions on X do we have the equality $(Z(X^*))^{com} = (Z^*(X))^{com}$?

At the present we do not know a complete solution to this problem. To illustrate it a bit deeper we give an example of a $C(K)$-module X for which there is no equality.

9.1.6. Consider $X = C(K) \oplus L^1(\mu)$, where $K = [0,1]$, and μ is the standard Lebesgue measure on K. We turn X into a $C(K)$-module by defining

$$f \cdot x = f \cdot (g,h) = (fg, fh)$$

whenever $f \in C(K)$ and $x = (g,h) \in X$ (i.e., $g \in C(K)$ and $h \in L^1(\mu)$). Let α be an arbitrary fixed non-zero linear multiplicative functional on $L^\infty(\mu)$ and let δ denote its restriction to $C(K)$. Let $(v,b) \in X^*$, where clearly $X^* = C(K)^* \oplus L^\infty(\mu)$. Finally, define $T(v,b) = (\alpha(b)\delta, 0)$. Obviously, $T \in L(X^*)$. Further, it is plain to see that each operator from $Z^*(X)$ is of the form $(v,b) \longrightarrow (fv, fb)$, where f is a fixed function from $C(K)$. This implies that $T \in Z^*(X)^{com}$. On the other hand, if P_δ is the band projection of $C(K)^*$ onto $\{\delta\}^{dd}$, then the operator $S : S(v,b) = (P_\delta v, 0)$ belongs to $Z(X^*)$, but $TS \neq ST$. ■

Even an extra assumption that $Z(X^*)^{com} = Z(X^*)$ does not "save" the situation; in this case there still exists an example of a Banach lattice X, for which $(Z^*(X))^{com} \neq Z(X^*)^{com}$. Our construction of this example depends on the following two ingredients.

(a) If Y is a Dedekind complete Banach lattice, and some $T \in L(Y)$ commutes with $Z(Y)$, then $T \in Z(Y)$. In other words, $Z(Y)^{com} = Z(Y)$. In particular, if X is any Banach lattice, then $Z(X^*)^{com} = Z(X^*)$, since X^* is Dedekind complete.

(b) There exists an infinite-dimensional Banach lattice X with the trivial center $Z(X)$, i.e., such that $T \in Z(X)$ iff $T = \lambda Id_X$ for some $\lambda \in \mathbf{C}$. †

To prove (a) notice that if B is a band in Y and P_B denotes the band projection on B, then, in view of our assumptions, $TP_B = P_B T$, which implies that $TB \subset B$, i.e., T is band preserving. Since B is arbitrary this implies that $T \in Z(Y)$.

Now let X be the Banach lattice mentioned in (b). Then obviously $(Z^*(X))^{com} = L(X^*)$. On the other hand, by (a), $Z(X^*)^{com} = Z(X^*)$ and we are done. ■

Not only for $C(K)$-modules, but even for Banach lattices, a general answer to Problem 2 is not clear. However, Theorem 9.11 shows that for a very large class of Banach lattice X the equality $Z(X^*)^{com} = (Z^*(X))^{com}$ does hold.

9.1.7. The inclusion in 9.1.4 suggests our last problem.

Problem 3. Let X be a $C(K)$-module or a BL. Is it true that $Z(X^*) = W(X^*)$?

It is worth mentioning that this problem was also raised in [HO1]. We give a positive solution to this problem in Theorem 9.12, which will be used at the end of this section to present a broad generalization of Sarason's and Gillespie's results.

---

† The first example of this kind was constructed by Goullet de Rugy [GdR]; we refer also to [Wic2], where this example and related matters are discussed in great detail.

9.2. Now, after this rather lengthy preliminary discussion, we are ready to set forth a detailed presentation.

Our next definition is an extension to $C(K)$-modules of the concept of orthomorphisms in the terminology of [AB], [Zaa] (or stabilizers in the terminology of [Mey], or unextending operators in the terminology of [AVK1]).

9.3. Definition. (1) *Let x be an arbitrary element of a $C(K)$-module X. The set*

$$supp(x) = \{t \in K : \forall\, f \in C(K) \quad (f(t) \neq 0) \Longrightarrow (fx \neq 0)\}$$

will be called the support set of the element x.

(2) *Let X, Y be two arbitrary $C(K)$-modules (with the same K). A linear operator $T : X \to Y$ is called $C(K)$-unextending (or a $C(K)$-orthomorphism) if $supp(Tx) \subset supp(x)$ for each $x \in X$.*

It is easy to see that $supp(x)$ is a closed subset of K for each element $x \in X$, and that $supp(x) = \emptyset$ if and only if $x = 0$. A straightforward verification also shows that, for each $x^* \in X^*$, the usual support of the measure $x^* \square x \in C(K)^*$ is a subset of $supp(x)$. This and Lemma 4.8(4) immediately imply the next result.

9.4. Lemma. *Let $x_1, x_2 \in X$ and $supp(x_1) \cap supp(x_2) = \emptyset$. Then the elements x_1, x_2 of the $C(K)$-module X are disjoint in X.*

9.5. Theorem. *Let X, Y be two $C(K)$-modules (with the same K), and let $T : X \to Y$ be a linear continuous operator. The following three conditions are equivalent.*

(i) *T is $C(K)$-unextending.*

(ii) $fx = 0 \Longrightarrow fTx = 0, \quad f \in C(K),\ x \in X.$

(iii) *T is $C(K)$-linear, i.e., $T(fx) = fTx,\ f \in C(K),\ x \in X$.*

Proof. The validity of implications (iii) $\Longrightarrow$ (ii) $\Longrightarrow$ (i) is trivial, and this is true even without the assumption of continuity of T.

Let us verify that (i) $\Longrightarrow$ (iii). Fix an $x \in X$ and introduce the linear operator $S : C(K) \longrightarrow Y$ defined by

$$Sf = fTx - T(fx).$$

Obviously S is continuous and $S(\mathbf{1}) = 0$. Note also that since T is $C(K)$-unextending, we have $supp(Sh) \subset supp(h)$ for each $h \in C(K)(*)$. (Since the space $C(K)$ is a $C(K)$-module, the Definition 9.3 of $supp(h)$ is applicable to $h \in C(K)$. Obviously it coincides here with the usual definition of the support set of a continuous function.) Let us verify that S is a d-homomorphism. Pick two arbitarary disjoint functions $f, g \in C(K)$. In view of $(*)$, we have $supp(Sf) \cap supp(Sg) = \emptyset$ and, therefore by Lemma 9.4, $Sf \text{ d } Sg$ in Y. Our claim that S is a d-homomorphism is proved. Therefore, by Theorem 6.6, $Sf_1 \in \Delta(Sf_2)$ for each $f_1 \in \Delta(f_2)$. Taking $f_2 = \mathbf{1}$, we get $Sf_1 \in \Delta(S\mathbf{1}) \equiv 0$, that is, $S \equiv 0$. In other words $fTx = Tfx$. ■

9.5.1. Remark. In the case when $X = Y$ an independent proof for Theorem 9.5 may be found in [Eva], Theorem 1. To get it, one only has to prove that the mapping $x \mapsto supp(x)$, $x \in X$, satisfies all the properties of the *support representation map* defined in [Eva], Definition 2.

Notice also that when $X = Y$ condition (iii) simply means that $T \in C(K)^{com}$, and the latter condition is obviously equivalent to the statement that $T \in Z(X)^{com}$. In other words, Theorem 9.5 gives a pleasant description of the commutant of the center.

9.5.2. Remark. The next example shows that the assumption of continuity of the operator T in Theorem 9.5 is essential. The same example shows also that if we omit the continuity assumption in the definition of the algebra $W(X^*)$, then Theorem 9.12 asserting that $W(X^*)$ coincides with $Z(X^*)$ does not hold any more.

9.6. Example. Let t_0 be a fixed point in $K = [0, 1]$ and let δ denote the Dirac measure concentrated at t_0.

(a) Letting $X = C(K)^*$, we will consider X as a $C(K)$-module with the usual multiplication of functions and measures. Further, let D denote the collection of all measures μ on K for each of which there exists an open neighborhood U of t_0 (depending on μ) satisfying $|\mu|(U \backslash \{t_0\}) = 0$. Clearly D is a linear proper subspace of X. It is plain to see that there exist a measure $\nu \in X$ and a function $f_0 \in C(K)$ such that $f_0(t_0) = 0$ and $f_0\nu \notin D$. Therefore we can find a discontinuous linear functional α on X satisfying the next two conditions: $\alpha(f_0\nu) \neq 0$ and $\alpha(\mu) = 0$ for all $\mu \in D$. Now we define an operator $T : X \to X$ by the following formula: $T\mu = \alpha(\mu)\delta$. We omit a routine verification that T satisfies conditions (i) and (ii) of Theorem 9.5 but, nevertheless, T is not $C(K)$-linear. Indeed, $f_0 T\nu = f_0 \alpha(\nu)\delta = 0$, though $T(f_0\nu) = \alpha(f_0\nu)\delta \neq 0$.

(b) Now let $X = C(K)$ and consider $X^* = C(K)^*$ as the conjugate module, i.e., in the sense of Definition 7.8. We introduce the same $T = \alpha \otimes \delta : X^* \to X^*$ as in (a). Clearly $T \notin Z(X^*)$ since T is discontinuous. On the other hand, it is easy to check that T maps each $\sigma(X^*, X)$-closed ideal of X^* into itself. ■

Later we will need the following characterization of the algebra $W(X^*)$.

9.7. Lemma. *Let X be a $C(K)$-module or a Banach lattice and $T \in L(X^*)$. The following conditions are equivalent.*

(1) $T \in W(X^*)$.

(2) $x^* \square x = 0 \Longrightarrow Tx^* \square x = 0$, *where* $x^* \in X^*$, $x \in X$.

Proof. (1) $\Longrightarrow$ (2). Let $T \in W(X^*)$ and let $x^* \square x = 0$. By Lemma 4.8, the functional x^* is orthogonal to $X(x)$. Therefore Tx^* is also orthogonal to $X(x)$, and thus $(Tx^*)\square x = 0$. Let us prove the converse implication (2) $\Rightarrow$ (1). Take an arbitrary $\sigma(X^*, X)$-closed ideal E in X^* and denote by E_0 its orthogonal complement in X. Clearly, E_0 is a closed ideal in X. Condition (2) means that if x^* is orthogonal to E_0, then Tx^* is likewise orthogonal to E_0, that is, $TE \subset E$. ■

The previous lemma and Theorem 9.5 imply the following proposition.

9.8. Proposition. *Let X be a $C(K)$-module or a Banach lattice. Then $W(X^*) \subset (Z^*(X))^{com}$.*

Proof. Let K_1 be the maximal ideal space of the algebra $Z(X)$. Then we can identify $Z(X)$ with $C(K_1)$. Further, we introduce in X^* the structure of a $C(K_1)$-module by the following formula: $(fx^*)(x) = x^*(fx)$, where $x^* \in X^*, x \in X$, and $f \in C(K_1) = Z(X)$. Let $T \in W(X^*)$. We want to verify that $TS^* = S^*T$ for each $S \in Z(X)$. In view of Theorem 9.5, it is enough to show that $S^*x^* = 0 (*) \Longrightarrow S^*Tx^* = 0$ for each $S \in Z(X)$ and $x^* \in X^*$. As we noticed in 9.1.1, $x^* \square Sx = (S^*x^*) \square x$ for any $x^* \in X^*, x \in X$, and therefore the equality $(*)$ implies that $x^* \square Sx = 0$. Consequently, by Lemma 9.7, $Tx^* \square Sx = 0$ and hence $(S^*Tx^*) \square x = 0$. Since x is arbitrary we can conclude that $S^*Tx^* = 0$. ■

Our next aim is to characterize $C(K)$-modules or Banach lattices with an order continuous norm in terms of the algebras $Z^*(X)$ and $Z(X^*)$. In the case of Banach lattices this result is known [Wic3]. For completeness we present an independent proof of it. Our proof depends on a theorem of Luxemburg and Zaanen (cited below), while the proof in [Wic3] avoids it.

9.9. Proposition. *Let X be a Banach lattice. The following two conditions are equivalent.*

(1) *X has an order continuous norm (this implies in particular, that X is Dedekind complete).*

(2) $Z^*(X) = Z(X^*)$.

Proof. (1) $\Longrightarrow$ (2). In view of the inclusion $Z^*(X) \subset Z(X^*)$ (see 9.1.1), we need to verify only the converse inclusion $Z(X^*) \subset Z^*(X)$. Take an arbitrary $T \in Z(X^*)$, and apply 9.1.1 to T^*. Then $T^* \in Z(X^{**})$ and thus $T^*(x^{**}) \in \lambda\Delta(x^{**})$ (#) for some positive λ.

Recall that order continuity of the norm in X implies that X is an order ideal in X^{**} (under the canonical embedding $X \to X^{**}$). Let $S := T^*|X$. It is plain to see that (#) implies that S maps X into itself and, moreover, $S \in Z(X)$. Clearly $S^* = T$, that is, $T \in Z^*(X)$.

(2) $\Longrightarrow$ (1). It is a well-known theorem due to Luxemburg–Zaanen that the norm in a Banach lattice X is order continuous if and only if each band in X^* is $\sigma(X^*, X)$-closed. (See [LZ1], Theorem 36.2 or [AB2], Theorem 9.1.) Now assume by way of contradiction that the norm of X is not order continuous. Therefore there exists a band $B \subset X^*$ that is not $\sigma(X^*, X)$-closed. Denote by P the band projection of X^* onto B^d. Clearly $P \in Z(X^*)$. Notice now that each $T^* \in Z^*(X)$ is $\sigma(X^*, X) - \sigma(X^*, X)$ continuous and therefore $(T^*)^{-1}(0)$ is $\sigma(X^*, X)$-closed. On the other hand, $P^{-1}(0) = B$ is not $\sigma(X^*, X)$-closed, and hence $P \notin Z^*(X)$. ■

9.10. Theorem. *Let X be a $C(K)$-module or a Banach lattice. The following conditions are equivalent.*

(1) *For each $x \in X$, the norm of the Banach lattice $X(x)$ is order continuous.*

(2) $Z^*(X) = Z(X^*)$.

Proof. If X is a Banach lattice, then (1) is just another way of saying that the norm in X is order continuous, and we may apply the previous proposition.

Let X be a $C(K)$-module. Take an arbitrary $T \in Z(X^*)$. It is obvious that if x^* is orthogonal to $X(x)$, then Tx^* is also orthogonal to $X(x)$. Let us have (1). Fix for a moment an $x \in X$ and consider the Banach lattice $X(x)$. Since $T|X(x)^* \in Z(X(x)^*)$ and the norm in $X(x)$ is order continuous, by the already proven part of the theorem we have $T|X(x)^* \in Z^*(X(x))$, which means that there exists an operator $S \in Z(X(x))$ such that $S^* = T|X(x)^*$. Thus $Sx \in \|T\|\Delta(x)$ and $(Tx^*)(x) = x^*(Sx)$ for each $x^* \in X^*$. Now stop treating x as a fixed element. Then the last equality means that S is a linear operator on X, $S \in Z(X)$ and $S^* = T$. That is, (1) $\Longrightarrow$ (2).

To prove the converse implication (2) $\Longrightarrow$ (1), assume that for some $x_1 \in X$, condition (1) does not hold. Therefore (again by the validity of the theorem for Banach lattices) there exists an operator $T_1 \in Z(X(x_1)^*)\backslash Z^*(X(x_1))$.

As follows from Theorem 7.6 (see ($\star\star$)), we can find an element $h \in L^\infty_{X(x_1)}$ such that $T_1 x_1^* \,\square\, y_1 = h\cdot(x_1^* \,\square\, y_1)$ for all $x_1^* \in X(x_1)^*$ and $y_1 \in X(x_1)$. Since $L^1_{X(x_1)}$ is a band in L^1_X, we may assume without loss of generality that $h \in L^\infty_X$. This clearly implies that the operator $T : X^* \to X^*$ defined by $Tx^* \,\square\, x = h \cdot (x^* \,\square\, x)$ $(x^* \in X^*,\ x \in X)$ belongs to $Z(X^*)\backslash Z^*(X)$, a contradiction. ■

Our next result describes a large class of vector lattices, for which we can obtain a positive solution to Problem 2. We precede it with a definition.

9.11.0. Definition. *We say that a Banach lattice X has a topologically rich center if for each $x \in X$ and for each $y \in X_x$ we can find a norm bounded sequence $\{f_n\} \in Z(X)$ such that $\|f_n x - y\| \to 0$.*

In other words, this means that Banach lattice structure on X coincides with that of the $Z(X)$-module. The most important examples of Banach lattices with topologically rich center are the Dedekind complete Banach lattices and the Banach lattices with a quasi-interior point.

9.11. Theorem. (1) *Let X be an arbitrary Banach lattice with a quasi-interior point. Then* $Z(X^*) = W(X^*) = Z(X^*)^{com} = (Z^*(X))^{com}$.

(2) *Let X be an arbitrary Banach lattice with a topologically rich center. Then* $Z(X^*) = (Z^*(X))^{com}$.

Proof. (1) The inclusion $W(X^*) \subset (Z^*(X))^{com}$ has been proved in Proposition 9.8. To verify the converse inclusion, let us take an arbitrary $T \in (Z^*(X))^{com}, x^* \in X^*$, $x \in X$, and assume that $x^* \,\square\, x = 0$. If x_0 is an arbitrary quasi-interior point in X, then $y = |x| + |x_0|$ is also a quasi-interior point, and there exists an operator $S \in Z(X)$ such that $x = Sy$. Consequently $(S^*x^*) \,\square\, y = x^* \,\square\, Sy = 0$, and therefore $S^*x^* = 0$, since $X(y) = X$. Thus,

$$(Tx^*) \,\square\, x = (Tx^*) \,\square\, Sy = (S^*Tx^*) \,\square\, y = (TS^*x^*) \,\square\, y = 0.$$

By Lemma 9.7, $T \in W(X^*)$, that is, $(Z^*(X))^{com} \subset W(X^*)$.

Because the equality $Z(X^*) = Z(X^*)^{com}$ is true for all Banach lattices (see 9.1.6(a)), and because the inclusion $Z(X^*) \subset W(X^*)$ is trivial (see 9.1.4), it only remains to prove that $W(X^*) \subset Z(X^*)$.

Fix some $T \in W(X^*)$ and assume that

$$\sum_{i=1}^{i=k} x_i^* \,\square\, x_i = 0 \tag{9.2}$$

for some $x_i^* \in X^*$, $x_i \in X$, $i = 1, 2, \ldots, k$, $k \in \mathbf{N}$. We claim that $\sum_i Tx_i^* \,\square\, x_i = 0$. To this end let us introduce $e = |x_0| + \sum_i |x_i|$. Since e is a quasi-interior element in X, we can find $S_i \in Z(X)$ such that $x_i = S_i e$ for each i. Condition (9.2) implies that $(\sum_i S_i^* x_i^*) \,\square\, e = 0$. Therefore by Lemma 9.7, $T(\sum_i S_i^* x_i^*) \,\square\, e = 0$. Finally, making use of the equality $W(X^*) = (Z^*(X))^{com}$ we get:

$$0 = T(\sum_i S_i^* x_i^*) \,\square\, e = (\sum_i S_i^* T x_i^*) \,\square\, e = \sum_i Tx_i^* \,\square\, S_i e = \sum_i Tx_i^* \,\square\, x_i.$$

In view of the claim just proved, the equality $U(x^* \square x) := (Tx^*) \square x, \quad x^* \in X^*, x \in X$ defines a linear continuous operator U on L^1_X. Since X is a Banach lattice with a quasi-interior point, it follows from Corollary 7.7.1 that the unit ball of $Z(X)$ is $\sigma(L^\infty_X, L^1_X)$-dense in the unit ball of L^∞_X. Since $T \in (Z^*(X))^{com}$ we have $Tf^* = f^*T$ for each $f \in Z(X)$. The last two facts imply that $Uh = hU$ for each $h \in L^\infty_X = Z(L^1_X)$. That is, we have proved that U commutes with each central operator on L^1_X. As we proved in 9.1.6(a), this implies that $U \in Z(L^1_X) = L^\infty_X$. Thus $U = h$ for some $h \in L^\infty_X$, i.e., $Tx^* \square x = hx^* \square x$ for all $x^* \in X^*$, $x \in X$. This is equivalent to the desired statement that $T \in Z(X^*) = L^\infty_X$.

(2) Take an arbitrary $T \in (Z^*(X))^{com}$ and let $x \in X$. As we know the space $X(x)^*$ can be identified with a band in X^*. Let P_x denote the corresponding band projection from X^* onto $X(x)^*$. The assumption that X has a topologically rich center clearly implies that $P_xT \in (Z^*(X(x)))^{com}$. But $X(x)$ has a quasi-interior point, and so, by part (1), the operator $P_xT \in Z(X(x)^*)$. Notice also that $Z(X(x)^*)) = P_xZ(X^*)$. To finish the proof it suffice to choose a net $\{x_\alpha\}$ in X such that the ideals $X(x_\alpha)$ are increasing by inclusion and their union is X. ■

9.11.1. Remark. (a) The equality $Z(X^*) = W(X^*)$ is also true for Banach lattices with a topologically rich center, and it is a special case of Theorem 9.12.

(b) We do not know if the assumption of the topological richness of the center is also necessary for the validity of the equality $Z(X^*) = (Z^*(X))^{com}$.

Our next step is to show that the previous result implies a theorem which can be considered as a dual version of Theorem 6.2.

9.12. Theorem. (Dual Bade's Theorem) *Let X be a $C(K)$-module or a Banach lattice. Then* $Z(X^*) = W(X^*)$.

Proof. In view of 9.1.4 it is enough to verify that $W(X^*) \subset Z(X^*)$. Let $T \in W(X^*)$ and fix an $x \in X$. The space $X(x)^*$ may be identified with $X^*/X(x)^\perp$, where $X(x)^\perp$ is the orthogonal complement of $X(x)$ in X^*. Since $T(X(x)^\perp) \subset X(x)^\perp$, a quotient map $T_1 : X(x)^* \longrightarrow X(x)^*$ is defined by the following formula:

$$T_1(x^* + X(x)^\perp) = Tx^* + X(x)^\perp.$$

Take now an element $x^* + X(x)^\perp \in X(x)^*$ and an element $y \in X(x)$ such that $(x^* + X(x)^\perp) \,\square\, y = 0$. Then $x^* \,\square\, y = 0$ (we consider $L^1_{X(x)}$ as a band in L^1_X), and therefore (by Lemma 9.7) $Tx^* \,\square\, y = 0$. This implies that $T_1(x^* + X(x)^\perp) \,\square\, y = 0$, and hence (again by Lemma 9.7) $T_1 \in W(X(x)^*)$. By the previous theorem, $T_1 \in Z(X(x)^*)$. In particular, for each $x^* \in X^*$, we have $|(Tx^*) \,\square\, x| = |T_1(x^* + X(x)^\perp) \,\square\, x| \leq \|T\| \cdot |x^* \,\square\, x|$. In view of Lemma 7.11(3), the last inequality is equivalent to the inclusion $Tx^* \in \|T\|\Delta(x^*)$. Since x^* and x are arbitrary we can conclude that $T \in Z(X^*)$. ■

9.13. In view of the analogy between Theorems 9.12 and 6.2, it is natural to call Theorem 9.12 a dual Bade theorem. However, it is worth pointing out that this analogy is not complete since here we, *a priori,* assume that the algebra $W(X^*)$ consists of continuous operators. If we agree to consider this difference insignificant, then the dual Bade theorem is stronger than Theorem 6.2, since in the latter we consider all ideals, while in the former we consider only w*-closed ones.

In the case of a Banach lattice X, this theorem (with the same reservation about continuity) improves a theorem due to Abramovich, Veksler and Koldunov [AVK1] which (when applied to X^*) states that $Z(X^*)$ consists of band preserving operators on X^*.

9.14. Let X be an operator $C(K)$-module. In Theorem 6.2, we have characterized $Z(X)$ in topological terms as the closure of $C(K)$ in the strong operator topology. Now we are going to present an analog of this statement for the algebra $Z(X^*)$.

9.14.1. As is well known, the algebra $L(X^*)$ is the conjugate space of the projective tensor product $X^* \widehat{\otimes} X$ under the following duality: $< T, x^* \otimes x >= (Tx^*)(x)$, where $T \in L(X^*)$, $x^* \in X^*$ and $x \in X$. We will consider on $L(X^*)$ the following two topologies: $\sigma_1 = \sigma(L(X^*), X^* \widehat{\otimes} X)$ and $\sigma_0 = \sigma(L(X^*), X^* \otimes X)$. It is known that these two topologies coincide on the unit ball of $L(X^*)$.

9.14.2. Identifying $Z(X^*)$ with L^∞_X, we will consider on $Z(X^*)$ the weak topology $\sigma(L^\infty_X, L^1_X)$.

9.14.3. If X is an operator $C(K)$-module, then X^* may also be considered as an operator $C(K)$-module by letting $(fx^*)(x) = x^*(fx)$ for $x^* \in X^*, x \in X, f \in C(K)$. Therefore we can assume that $C(K)$ is embedded in $L(X^*)$.

9.15. Theorem. *Let X be an operator $C(K)$-module. Then*

(1) *The algebra $Z(X^*)$ coincides with the closure of the algebra $C(K)$ in either of the topologies σ_0 or σ_1.*

(2) *The closed unit ball of the algebra $Z(X^*)$ coincides with the closure of the unit ball of $C(K)$ in either of the topologies σ_0 or σ_1.*

(3) *All three topologies σ_0, σ_1 and $\sigma(L^\infty_X, L^1_X)$ coincide on $Z(X^*)$.*

Proof. Let $T \in Z(X^*)$ and let $l = \sum_{k=1}^{\infty} x_k^* \otimes x_k$ be an element in $X^* \hat{\otimes} X$ such that l is orthogonal to the subspace $C(K)$ of $L(X^*)$. Thus, for each $f \in C(K)$ we have

$$0 = l(f) = \langle \sum_k x_k^* \otimes x_k, f \rangle = \sum_k f^* x_k^*(x_k) = \langle \sum_k x_k^* \square\, x_k, f \rangle. \tag{9.3}$$

Since f is arbitrary, (9.3) means that the measure $\mu = \sum_k x_k^* \square\, x_k$ in L^1_X is zero. As $T \in Z(X^*) = L^\infty_X = (L^1_X)^*$, we have $< T, \mu >= 0$, and in view of $(\star)$ in 7.6 this gives $\sum_k T x_k^* \square\, x_k = 0$. Substituting T for f in (9.3) we obtain that $l(T) = 0$. Consequently, $T \in \sigma_1\text{-cl}(C(K))$. So certainly, $T \in \sigma_0\text{-cl}(C(K))$.

Now let T be an arbitrary operator from $\sigma_0\text{-cl}(C(K))$. It is easy to see that $T \in C(K)^{com}$ in $L(X^*)$. Fix an arbitrary $x^* \in X^*$ and $x \in X$ satisfying $x^* \square\, x = 0$. Then obviously, $f^* x^* \square x = 0$ for each $f \in C(K)$. In view of the duality between $X^* \otimes X$ and $L(X^*)$ the last statement means that the tensor $f^* x^* \otimes x$ is orthogonal to each $g \in C(K) \subset L(X^*)$, and therefore has to be orthogonal to T, i.e., $(Tf^*x^*)(x) = (f^*Tx^*)(x) = 0$. This implies that $Tx^* \square\, x = 0$. Therefore by Lemma 9.7, the operator T belongs to $W(X^*)$, and finally, by Theorem 9.12, $T \in Z(X^*)$. This proves (1).

Obviously, in view of Corollary 7.7.1, (3) implies (2).

To prove (3) notice that there is a continuous mapping of $X^* \widehat{\otimes} X$ into L^1_X, given by the following formula $\sum_{k=1}^{\infty} x_k^* \otimes x_k \mapsto \sum_{k=1}^{\infty} x_k^* \square\, x_k$. This implies that on the space L^∞_X we have $\sigma_0 \le \sigma_1 \le \sigma(L^\infty_X, L^1_X)$. On the other hand, by Theorem 8.1, each element μ of L^1_X can be represented as $x^* \square\, x$ for some $x^* \in X^*$ and $x \in X$. Therefore $\sigma_0 \equiv \sigma(L^\infty_X, L^1_X)$ on L^∞_X. ∎

9.16. Now we are ready to present Theorem 9.17, one of the major results of this section. In this theorem we describe the closure of any unital subalgebra $\mathcal{A}$ of $Z(X^*)$ in the weak* operator topology $\sigma_0 = \sigma(L(X^*), X^* \otimes X)$, where X is an arbitrary $C(K)$-module. A similar result for subalgebras of $L(X)$ is stated in Theorem 9.18. Both these theorems are broad generalizations of well known results of W. Bade, D. Sarason [Sar] and T. Gillespie [Gil]. The factorization Theorem 8.1 and Theorems 9.12 and 6.2 are the principal tools which have allowed us to adapt the technique used in [Gil], where subalgebras of complete Boolean algebras of projections are considered. We will denote by $\overline{\mathcal{A}}$ the $\sigma(L(X^*), X^* \otimes X)$-closure of $\mathcal{A} \subset L(X)$.

9.17. Theorem. *Let X be an arbitrary $C(K)$-module and $\mathcal{A}$ be an unital subalgebra of $Z(X^*)$. An operator $T \in L(X^*)$ belongs to $\overline{\mathcal{A}}$ if and only if T leaves invariant each σ_0-closed subspace of X^* which is invariant under each $A \in \mathcal{A}$. (In short: $TV \subseteq V$ for each σ_0-closed $V \in Lat\,\mathcal{A}$.)*

Proof. The "only if" part is trivial.

To prove the "if" part assume that T satisfies the conclusion of the theorem. Obviously, T leaves invariant each σ_0-closed ideal in X^*, and consequently Theorem 9.12 implies that $T \in Z(X^*)$. Now observe that by the definition of the topology $\sigma_0 = \sigma(L(X^*), X^* \otimes X)$ we obviously have the following equivalence :

$$T \in \overline{\mathcal{A}} \Longleftrightarrow \sum_{k=1}^{n} < Tx_k^*, x_k > = 0, \tag{9.4}$$

whenever

$$\sum_{k=1}^{n} < Ax_k^*, x_k > = 0 \tag{9.5}$$

for some $x_k^* \in X^*, x_k \in X, n \in \mathbf{N}$ and all $A \in \mathcal{A}$. Fix some $x_k^* \in X^*$ and $x_k \in X$ satisfying (9.5). By Theorem 8.1 there exist $x^* \in X^*, x \in X$ such that

$$\sum_k x_k^* \square x_k = x^* \square x. \tag{9.6}$$

Therefore (9.5) implies that

$$< Ax^*, x > = 0 \quad \forall A \in \mathcal{A}. \tag{9.7}$$

Consider the σ_0-closed subspace V of X^* generated by $\{Ax^* : A \in \mathcal{A}\}$. Obviously $AV \subseteq V$, and therefore $TV \subseteq V$. Since $\mathcal{A}$ is unital, the element $x^* \in V$. Clearly (9.7) implies that V is orthogonal to x. Since $TV \subseteq V$ and $x^* \in V$, we conclude that $Tx^* \in V$, and therefore $< Tx^*, x >= 0$. By virtue of (9.6) this means that $\sum_k < Tx_k^*, x_k >= 0$. That is, in view of (9.4), we have $T \in \overline{\mathcal{A}}$. ■

The proof of the next theorem is almost identical and will be omitted. The only difference is that instead of Theorem 9.12, one should apply Theorem 6.2.

9.18. Theorem. *Let X be an operator $C(K)$-module and $\mathcal{A}$ a unital subalgebra of $Z(X)$. A linear operator $T : X \to X$ (which is not assumed a priori to be continuous) belongs to the closure of $\mathcal{A}$ in the weak operator topology on $L(X)$ if and only if T leaves invariant each closed subspace of X which is invariant under each $A \in \mathcal{A}$. (In short: $TV \subseteq V$ for each $V \in Lat\,\mathcal{A}$; in particular, $\mathcal{A}$ is reflexive (see 6.2.3), i.e., $\mathcal{A} = AlgLat\mathcal{A}$.)*

10. Operators conjugate to d-homomorphisms

10.1. Let $T \in dh(X,Y)$, where X and Y are Banach lattices. Then the conjugate operators $T^* : Y^* \to X^*$ and $T^{**} : X^{**} \to Y^{**}$ have several "good" properties; for example, (i) T^* is a product of an interval preserving operator (a Maharam operator in another terminology) and an operator from $Z(X^*)$, and (ii) T^{**} is an order continuous d-homomorphism. We will show that some analogs of these results still hold for operators between $C(K)$-modules.

The properties of the operator T^{**} will be extensively used in Sections 12 and 13 in connection with the study of the spectrum of d-homomorphisms.

10.2. The following notations will be fixed throughout this section. The letters X and Y will denote either two Banach lattices or a $C(K_1)$-module and a $C(K_2)$-module, respectively. By T we denote an operator in $dh(X,Y)$. The conjugate space X^* is always viewed as the $Z(X^*)$-module described in Definition 7.8 and the paragraph thereafter. This module generates in the same way the space $L^1_{X^*}$ and the center $Z(X^{**})$ of the second conjugate space. Recall that $Z(X^{**}) = L^\infty_{X^*} = (L^1_{X^*})^*$. Similarly, we introduce $L^1_{Y^*}$ and $Z(Y^{**}) = L^\infty_{Y^*}$.

10.3. Let P_1 denote the band projection of L^1_Y onto the band generated by all the elements of the form $y^* \square Tx$, where $y^* \in Y^*$ and $x \in X$.

Because there is a natural isomorphism between the bands of L^1_Y and those of $(L^1_Y)^* (= L^\infty_Y = Z(Y^*))$ we can consider P_1 as an element of the algebra $Z(Y^*)$. It is easy to verify that P_1 is the least projection in $Z(Y^*)$ for which $T^*P_1 = T^*$. The support set of P_1 will be denoted by E_{T^*}.

Similarly, P_2 will denote the band projection of $L^1_{Y^*}$ onto the band generated by all elements of the form $T^{**}x^{**} \square y^*$, where $x^{**} \in X^{**}$ and $y^* \in Y^*$. In this case, P_2 is the least projection in $Z(Y^{**})$ for which $P_2T^{**} = T^{**}$. The support set of P_2 will be denoted by $E_{T^{**}}$.

As usual, $Q_1 := Q(Z(X^*))$ denotes the Stone (or, equivalently, Gelfand) space of $Z(X^*)$, and similarly $Q_2 := Q(Z(X^{**}))$.

10.4. Theorem. *There exists a unique algebraic homomorphism*

$$\beta_{T^*} : Z(X^*) \longrightarrow P_1 Z(Y^*)$$

such that $\beta_{T^*}(\mathbf{1}) = P_1$ *and* $HT^* = T^*\beta_{T^*}(H)$ for all $H \in Z(X^*)$.

Proof. We divide our proof into several steps. In the first three steps X is a $C(K_1)$-module and Y is a $C(K_2)$-module.

Step 1. Let $T^*G = 0$ for some $G \in P_1 Z(Y^*)$. We claim that $G = 0$. Indeed, for an arbitrary $y^* \in Y^*$ and $x \in X$ we have $0 = < T^*Gy^*, x > = < Gy^*, Tx >$. This means that $G(y^* \square Tx) = 0$. The definition of P_1 implies that $G = P_1 G = 0$.

Step 2. Let Y_1 be the closed ideal in Y generated by $T(X)$. Then 4.15 and the definition of P_1 imply that $L^1_{Y_1} = P_1 L^1_Y$.

Step 3. By Theorem 6.7, there exists a unique algebraic homomorphism γ from $Z(X)$ into $Z(Y_1)$ such that $Tf = \gamma(f)T$ for all $f \in Z(X)$.† Let $H \in Z(X^*), \|H\| \leq 1$. By Theorem 9.15, there exists a net $\{f_\alpha\}$ in the unit ball of $C(K_1)$ converging to H in the topology $\sigma(L(X^*), X^* \otimes X)$. Without loss of generality we can assume that the net $\{\gamma(f_\alpha)\}$ converges to some element $G \in Z(Y_1^*)$ in the topology $\sigma(L(Y_1^*), Y_1^* \otimes Y_1)$. In view of Step 2 we can assume that $G \in P_1 Z(Y^*)$. Now for an arbitrary $y^* \in Y^*$ and $x \in X$ we have

$$\begin{aligned}(HT^*y^*)(x) &= \lim_\alpha y^*(Tf_\alpha x) = \lim_\alpha y^*(\gamma(f_\alpha)Tx) = \\ &= \lim_\alpha (\gamma(f_\alpha)^* y^*)(Tx) = (T^*Gy^*)(x),\end{aligned}$$

whence $HT^* = T^*G$.

Step 1 guarantees that the operator $G \in P_1 Z(Y^*)$ is uniquely determined, that is, the introduced mapping $\beta_{T^*} : H \longrightarrow G$ is well defined. It is plain to see that this mapping is an algebraic homomorphism sending $\mathbf{1}$ to P_1.

Step 4. Let now X and Y be Banach lattices. For each $x \in X$, the annihilator $X(x)^\perp$ is a band in X^*. Its disjoint complement $\{X(x)^\perp\}^d$ may be identified with

---

† Certainly this implies that $f^*T^* = T^*\gamma(f)^*$ for all $f \in Z(X)$, where the conjugate operator means an operator between Y_1^* and X^*, whereas a more general identity we want to prove is for all operators $H \in Z(X^*)$, and for the conjugate $T^* : Y^* \rightarrow X^*$.

$X(x)^*$. We denote by R_x the projection of X^* onto this band $X(x)^*$. By P_x we denote a similar band projection of Y^* onto $Y(Tx)^*$. Since $T(X(x)) \subset Y(Tx)$, and since the unit ball of the algebra $Z(X(x)^*)$ coincides with the $\sigma(L(X(x)^*), X(x)^* \otimes X(x))$-closure of the unit ball of $Z^*(X(x))$, Step 3 shows that there exists a homomorphism $\beta_x : R_x Z(X^*) \to P_1 P_x Z(Y^*)$ such that

$$\beta_x(\mathbf{1}) = P_1 P_x$$

and

$$HT^* P_x = T^* \beta_x(R_x H), \quad H \in Z(X^*).$$

Pick a net $\{x_\alpha\} \subset X$ such that the ideals $X(x_\alpha)$ increase (by inclusion) and their union $\bigcup_\alpha X(x_\alpha) = X$. For each $H \in Z(X^*)$, the net $\beta_{x_\alpha}(R_{x_\alpha} H)$ has a limit point G in the topology $\sigma(L(Y^*), Y^* \otimes Y)$.

Hence, $HT^* = T^*G$. Letting $\beta_{T^*}(H) = G$, we get the desired homomorphism. ■

10.5. Theorem. *Let $\varphi_{T^*} : E_{T^*} \to Q_1$ be the unique mapping corresponding to the homomorphism β_{T^*}, that is, $\beta_{T^*}(H) = H \circ \varphi_{T^*}$. Then φ_{T^*} is an open map.*

Proof. In view of 3.7(2) it is enough to verify that the homomorphism β_{T^*} is order continuous as an operator from the Banach lattice $Z(X^*)$ into the Banach lattice $P_1 Z(Y^*)$. Let $H_\alpha \downarrow 0$ in $Z(X^*)$ and assume that $\inf_\alpha \beta_{T^*}(H_\alpha) = G \neq 0$. Then $G = P_1 G$ and there are $y^* \in Y^*$ and $x \in X$ such that $(Gy^*)(Tx) \neq 0$. On the other hand, the functional $H \mapsto (HT^*y^*)(x)$ is clearly order continuous on $Z(X^*)$. Therefore,

$$0 = \lim_\alpha (H_\alpha T^* y^*)(x) = \lim_\alpha (T^* \beta_{T^*}(H_\alpha) y^*(x) = (Gy^*)(Tx) \neq 0,$$

a contradiction. ■

Since the proofs of the following two results about the second conjugate T^{**} are more or less identical to the previous proofs, we confine ourselves only to the corresponding statements.

10.6. Theorem. *There exists a unique algebraic homomorphism $\beta_{T^{**}}$ from the center $Z(X^{**})$ into $P_2 Z(Y^{**})$ such that $\beta_{T^{**}}(\mathbf{1}) = P_2$ and $T^{**}H = \beta_{T^{**}}(H) T^{**}$ for each $H \in Z(X^{**})$.*

10.7. Theorem. *Let* $\varphi_{T^{**}} : E_{T^{**}} \to Q_2$ *be the unique mapping corresponding to the homomorphism* $\beta_{T^{**}}$*, that is,* $\beta_{T^{**}}(H) = H \circ \varphi_{T^{**}}$*. Then* $\psi_{T^{**}}$ *is an open map.*

10.8. Remark. Theorems 10.6 and 10.7 above show that T^{**} is an order continuous homomorphism. The precise definition of order continuity for d-homomorphisms between $C(K)$-modules is given in Definition 11.27.

11. Independence of d-homomorphisms

In this section, we introduce and investigate a new concept of independence for the d-homomorphisms, a concept which is of interest in its own right and which will be extensively used in the next section devoted to the spectral properties of d-endomorphisms. In the case of Dedekind complete vector lattices, the independence of operators means precisely their disjointness in the Dedekind complete space of all regular operators.

11.1. Let X and Y be vector lattices, $L_b(X,Y)$ the space of all order bounded operators from X into Y. If Y is Dedekind complete, then by a well known Kantorovich theorem the space $L_b(X,Y)$ coincides with the space $L_r(X,Y)$ of all regular operators.

If Y is a vector lattice which is not Dedekind complete, then (as earlier in Section 6.8.1) we assume that Y is embedded in $\widehat{Y}$ and J is the operator of the corresponding canonical embedding.

11.2. Definition. *Operators $T_1, T_2 \in L_b(X,Y)$ are called independent (in symbols $T_1 \,\delta\, T_2$) if operators JT_1 and JT_2 are disjoint in the Dedekind complete vector lattice $L_b(X,\widehat{Y}) = L_r(X,\widehat{Y})$.*

Thus, $T_1 \,\delta\, T_2$ if and only if $JT_1 \,\mathrm{d}\, JT_2$ in $L_b(X,\widehat{Y})$. If Y is Dedekind complete, the introduced independence "δ" coincides with the usual disjointness "d". We write $T_1 \,\overline{\delta}\, T_2$, provided operators T_1 and T_2 are not independent.

11.3. Lemma. *For $T_1, T_2 \in L_b(X,Y)$ the following two conditions are equivalent.*

(1) *$T_1 \,\delta\, T_2$.*

(2) *$(T_1|X_x)\,\delta\,(T_2|X_x)$ for each $x \in X$.*

Proof. Without loss of generality, we may assume that $Y = \widehat{Y}$. Therefore, the independence "δ" in (1) and (2) is simply the classical disjointness. However, for disjointness the equivalence (1) $\Longleftrightarrow$ (2) is a direct consequence of the famous Kantorovich formula for evaluation of the infimum of two operators. (See [AB1], p.12; or [Vul], p.210.) ∎

11.4. Lemma. *Let X be a vector lattice with a strong unit x_0, Y be an (r_u)-complete vector lattice and $T_1, T_2 \in dh(X,Y)$. The following conditions are equivalent.*

(1) $T_1 \,\overline{\delta}\, T_2$.

(2) *There exists $T \in dh(X,Y)$ such that $0 < |JT| \le |JT_1| \wedge |JT_2|$.*

(3) *There exist $g_i \in Z(Y_{T_i x_0})$ $(i = 1,2)$ such that $g_1 T_1 = g_2 T_2 \ne 0$.*

Proof. Implications (3) $\Longrightarrow$ (2) $\Longrightarrow$ (1) are obvious. To prove (1) $\Longrightarrow$ (3), let us introduce $S = |JT_1| \wedge |JT_2| : X \to \widehat{Y}$ and let $y_0 := Sx_0$. Clearly, $S \in dh_+(X, \widehat{Y})$ and, in view of (1), $S > 0$. Therefore $y_0 > 0$, since x_0 is a strong unit. The following inequality is also obvious: $y_0 \le |JT_1 x_0| \wedge |JT_2 x_0| = |T_1 x_0| \wedge |T_2 x_0|$.

By Theorem 3.3 there exist $h_1, h_2 \in Z(\hat{Y})$ such that

$$(a) \qquad S = h_1 JT_1 = h_2 JT_2.$$

On the other hand, Theorem 6.7 implies that there exist three maps

$$\psi_{JT_1} : E_{JT_1} \to K, \quad \psi_{JT_2} : E_{JT_2} \to K, \quad \text{and} \quad \psi_S : E_S \to K,$$

where K is the Gelfand space of $Z(X)$, such that

$$(b) \qquad JT_k fx = (f \circ \psi_{JT_k}) JT_k x, \quad (k = 1,2) \quad \text{and} \quad Sfx = (f \circ \psi_S) Sx$$

for each $x \in X$ and $f \in Z(X)$. From (a) and (b) it follows immediately that on E_S

$$(c) \qquad \psi_{JT_1} = \psi_{JT_2} = \psi_S.$$

Now let us fix an arbitrary $y \in Y$ such that $0 < y \le y_0$. Then we can find an element $h \in Z(\hat{Y})$ satisfying $y = hy_0$, and consequently $hSx_0 = y_0$. Since Y is (r_u)-complete, there exist $g_k \in Z(Y_{T_k x_0})(k = 1,2)$ such that $y = g_1 T_1 x_0 = g_2 T_2 x_0$. Operators $g_1 T_1$ and $g_2 T_2$ map X into Y and both are nonzero because $y \ne 0$.

In view of (c) and of the equality $g_1 T_1 x_0 = g_2 T_2 x_0 = hSx_0$ it follows that the operators $g_1 T_1, g_2 T_2$ and hS coincide as operators from X into $\hat{Y}$. This implies that $g_1 T_1$ and $g_2 T_2$ coincide as operators from X into Y. ■

The previous two lemmas imply the next proposition.

11.5. Proposition. *For $T_1, T_2 \in dh(X,Y)$, where X is a vector lattice and Y is an (r_u)-complete vector lattice, the following two conditions are equivalent.*

(1) $T_1 \,\overline{\delta}\, T_2$.

(2) *There exist $x \in X$, $g_1 \in Z(Y_{T_1 x})$, and $g_2 \in Z(Y_{T_2 x})$ such that* $g_1 T_1 | X_x = g_2 T_2 | X_x \neq 0$.

11.6. Our next step is to find a reasonable extension to $C(K)$-modules of the definition of the independence of d-homomorphisms. The following discussion provides some explanations as to what type of an extension should be called reasonable.

Let X be a $C(K_1)$-module, Y be a $C(K_2)$-module and $Z(X)$ and $Z(Y)$ be their centers. Assume also that X and Y are simultaneously Banach lattices whose centers coincide with $Z(X)$ and $Z(Y)$ respectively. If the definition of independence is reasonable (appropriate, meaningful, right, etc), then, under the conditions imposed, it should, of course, be equivalent to Definition 11.2.

In a more general situation of arbitrary $C(K)$-modules our requirements as to what should be called reasonable are given in the following three conditions.

(I_1) If T_1 and T_2 are independent, then their restrictions $T_1|X'$, $T_2|X'$ to each closed ideal X' in X should remain independent.

(I_2) Let Y' be a closed ideal in Y such that $T_1(X) \subset Y'$ and $T_2(X) \subset Y'$. Then d-homomorphisms $T_1, T_2 : X \to Y$ are independent if and only if $T_1, T_2 : X \to Y'$ are independent.

(I_3) If $X = X(x)$ and $Y = Y(y)$ for some $x \in X$ and $y \in Y$, then $T_1, T_2 : X \longrightarrow Y$ are independent if and only if T_1 and T_2 are independent as operators from the Banach lattice $X(x)$ into the Banach lattice $Y(y)$.

The conditions (I_1) – (I_3) are already sufficient to have interesting consequences.

11.7. Definition. *Let $T \in dh(X,Y)$, where X is a $C(K_1)$-module and Y is a $C(K_2)$-module. For any $x \in X$, introduce $T^x \in dh(C(K_1),Y)$ by the following formula: $T^x f = T(fx)$, $f \in C(K_1)$.*

It is plain to see that $T^x \neq 0$ provided $T^x\mathbf{1}(= Tx) \neq 0$. Indeed, if $Tx = 0$, then, applying Theorem 6.7 to an arbitrary $f \in C(K_1)$, we obtain $T^x f = T(fx) =$

$\gamma(f)Tx = 0$.

11.7.1. Lemma. *Let X be a $C(K_1)$-module, Y be a $C(K_2)$-module and let $x_0 \in X, y_0 \in Y$ be such elements that $X = X(x_0), Y = Y(y_0)$. Therefore X and Y may be considered as Banach lattices, and consequently, the independence of operators is defined. For $T_1, T_2 \in dh(X,Y)$, the following conditions are equivalent.*

(1) $T_1 \ \delta \ T_2$.

(2) $T_1^{x_1} \ \delta \ T_2^{x_2}$ *for all* $x_1, x_2 \in X$.

Proof. (2) $\Longrightarrow$ (1). Assume, contrary to what we claim, that $T_1 \ \overline{\delta} \ T_2$. Then by Proposition 11.5, there exist $x \in X$ and $g_i \in Z(Y_{T_i x})$ $(i = 1, 2)$, such that $g_1 T_1 | X_x = g_2 T_2 | X_x \neq 0$. Whence $g_1 T_1^x = g_2 T_2^x \neq 0$, and hence, again by Proposition 11.5, $T_1^x \ \overline{\delta} \ T_2^x$.

(1) $\Longrightarrow$ (2). Assume that $T_1^{x_1} \ \overline{\delta} \ T_2^{x_2}$ for some $x_1, x_2 \in X$. Without loss of generality, we can assume that $|x_i| \leq x_0$ (otherwise, we replace x_0 by $|x_0| + |x_1| + |x_2|$), and therefore, $|T_i^{x_i}| \leq |T_i^{x_0}|$ $(i = 1, 2)$. Consequently $T_1^{x_0} \ \overline{\delta} \ T_2^{x_0}$, and by Lemma 11.4 we can find $g_i \in Z(Y_{T_i x_0})$ $(i = 1, 2)$ such that $g_1 T_1^{x_0} = g_2 T_2^{x_0} \neq 0$. In other words, $g_1 T_1(f x_0) = g_2 T_2(f x_0)$ for all $f \in Z(X)$. Since $\{f x_0 : f \in Z(X)\}$ is dense in X, we have that $g_1 T_1 = g_2 T_2$, that is, $T_1 \ \overline{\delta} \ T_2$. ■

The foregoing discussions in 11.6 and 11.7 justify the next definition.

11.8. Definition. *Let $T_1, T_2 \in dh(X,Y)$, where X is a $C(K_1)$-module and Y is a $C(K_2)$-module. We call T_1 and T_2 dependent (in symbols: $T_1 \ \overline{\delta} \ T_2$) if there are $x_1, x_2 \in X$, $g_1 \in Z(Y_{T_1 x_1})$ and $g_2 \in Z(Y_{T_2 x_2})$ such that $g_1 T_1^{x_1} = g_2 T_2^{x_2} \neq 0$.*

If T_1, T_2 are not dependent, they are called **independent** and, in this case, we write $T_1 \ \delta \ T_2$. It should be immediate that the above defined independence satisfies the conditions $(I_1) - (I_3)$.

11.9. Later, in Sections 12 and 13, we will show that the introduced concept of independence plays an important role in studying the spectrum of d-homomorphisms on $C(K)$-modules and Banach lattices. For example, in Theorem 12.11 we will prove that pairwise independence of different powers of any d-homomorphism $T \in dh(X)$ implies that the spectrum $\sigma(T)$ is rotation invariant. To prove this and other results some preparatory work on independence is needed. The following three paragraphs summarize what will be achieved in this section.

(a) We introduce a class of **nice** d-homomorphisms and with each nice d-homomorphism T we associate in a unique way some continuous mapping φ_T. Then we show that independence $T_1 \,\delta\, T_2$ may be expressed in terms of φ_{T_1} and φ_{T_2}.

(b) We introduce a class of order complete $C(K)$-modules, a class which is rather large and has many good properties. For example, all d-homomorphisms on such modules are nice and, as a result, we are able to find some workable conditions of their independence.

(c) For arbitrary d-homomorphisms T_1, T_2 on a $C(K)$-module X we find a criterion of independence of operators T_1 and $T = T_1T_2$ in terms of mappings $\varphi_{T^{**}}$ and $\varphi_{T_1^{**}}$. Though this criterion is of a technical nature it turns out to be very useful in the proofs of the main results of Sections 12 and 13.

11.10. Definition. *Let $T \in dh(X,Y)$, where X is an arbitrary $C(K_1)$-module and Y is a $C(K_2)$-module. We say that T is nice if for any $f \in Z(X)$ there exists $g \in Z(Y)$ such that $Tf = gT$. The collection of all nice d-homomorphisms will be denoted by $dh_n(X,Y)$.*

If T is a nice d-endomorphism on X, then we simply write $T \in dh_n(X)$ instead of $T \in dh_n(X,X)$.

At this point it seems appropriate to warn the reader against a possible confusion. Indeed, in view of Theorem 6.7 there may be a temptation to say that each d-homomorphism is nice. However, it is not true because Theorem 6.7 only states that for each $f \in Z(X)$ there exists $g = \gamma(f) \in Z(Y_{TX})$ (but not, in general, in $Z(Y)$), such that $Tf = gT$.

The following simple example illustrates this. Consider $C[0,1]$ as a $C[0,1]$-module, and let

$$Tf(t) = \begin{cases} tf(\sin \frac{1}{t}) & \text{if } t \neq 0 \\ 0 & \text{if } t = 0. \end{cases}$$

Clearly T is a d-homomorphism on $C[0,1]$. Assume, contrary to what we claim, that T is nice, and take $f_0(t) \equiv t \in C[0,1]$. Hence there would exist a function $g_0 \in C[0,1]$, such that $Tf_0 = g_0T$. Applying each of these two equal operators to the function $\mathbf{1} \in C[0,1]$ we get $t\sin\frac{1}{t} = g_0(t)t$, whence $g_0(t) = \sin\frac{1}{t} \in C[0,1]$, a contradiction.

Nevertheless, if T is an "onto" d-homomorphism, then Theorem 6.7 does allow us to prove that T is nice.

11.11. The next condition introduces a class of rather important modules.

We say that a $C(K)$-module X satisfies a condition $(\diamond)$, provided X is an operator module and $Z(X) = C(K)$.

11.11.1. Let X (resp. Y) be a $C(K_1)$-module (resp. a $C(K_2)$-module) satisfying $(\diamond)$.

Take now an arbitrary $T \in dh_n(X,Y)$, and let $J_T = \{g \in C(K_2) : gT = 0\}$. Obviously J_T is an ideal in $C(K_2)$. We denote by E_T its zero set. Therefore $C(K_2)/J_T$ may be identified with $C(E_T)$. Since for each $f \in C(K_1)$ there exists $g \in C(K_2)$ such that $Tf = gT$, and since for $f = 0$ the corresponding set of all such g's is J_T, we see that T generates in a unique way an algebraic homomorphism $\gamma_T : C(K_1) \to C(E_T)$. Accordingly, there exists a continuous mapping $\varphi_T : E_T \to K_1$ associated with γ_T, that is, $\gamma_T(f) = f \circ \varphi_T$ for all $f \in C(K_1)$. It is immediate now that if $Tf = gT$ for some $f \in C(K_1)$ and $g \in C(K_2)$, then $g(t) = (f \circ \varphi_T)(t)$ for each $t \in E_T$. This mapping φ_T will be used throughout the whole section.

It is easily seen that if $y = hTx \neq 0$ for some $x \in X$ and some $h \in Z(Y_{Tx})$, then $supp(y) \subset E_T$. Indeed, let $t_0 \in supp(y)$. Then for each $g \in J_T$ we necessarily have $g(t_0) = 0$, since otherwise $gy \neq 0$ contrary to $gy = ghTx = hgTx = 0$. That is, $t_0 \in E_T$.

11.12. Proposition. *Let $S, T \in dh_n(X,Y)$, where modules X and Y satisfy $(\diamond)$. If $S\,\bar{\delta}\,T$, then there exists a non-zero $y \in Y$ such that $supp(y) \subset \{q \in E_S \cap E_T : \varphi_S(q) = \varphi_T(q)\}$. Moreover, if S and T are "onto", then the converse conclusion is also true.*

Proof. Indeed, by Definition 11.8, $g_1S^{x_1} = g_2T^{x_2} \neq 0$ for some $x_1, x_2 \in X$, $g_1 \in Z(Y_{Sx_1})$, $g_2 \in Z(Y_{Tx_2})$, and we can take $y = g_1Sx_1 (= g_2Tx_2)$. As we noticed after Definition 11.7, necessarily $y \neq 0$. Notice also that the discussion preceding this proposition clearly implies that $supp(y) \subset \{q \in E_S \cap E_T\}$.

It remains to verify that $\varphi_S(q) = \varphi_T(q)$ for all $q \in supp(y)$. Assume, contrary to what we claim, that there exists a point $q_0 \in supp(y)$, such that $\varphi_S(q_0) \neq \varphi_T(q_0)$. Therefore, there exists some $f \in C(K_1)$, such that $f(\varphi_S(q_0)) \neq f(\varphi_T(q_0))$, that is, $((\gamma_S - \gamma_T)(f))(q_0) = (f \circ \varphi_S - f \circ \varphi_T)(q_0) \neq 0$. Denote by g_S (resp. g_T) an arbitrary extension of $\gamma_S(f)$ (resp. $\gamma_T(f)$) to an element of $C(K_2)$. Since $q_0 \in supp(y)$ and $(g_S - g_T)(q_0) \neq 0$ we can conclude that $(g_S - g_T) \cdot y \neq 0$. On the other hand we have:

$$\begin{aligned}0 &= (g_1 S^{x_1} - g_2 T^{x_2})(f) = g_1 S^{x_1} f - g_2 T^{x_2} f = g_1 S f x_1 - g_2 T f x_2 \\ &= g_1 g_S S x_1 - g_2 g_T T x_2 = g_S g_1 S x_1 - g_T g_2 T x_2 = g_S \cdot y - g_T \cdot y \\ &= (g_S - g_T) \cdot y,\end{aligned}$$

a contradiction.

Assume now that S and T are "onto." Consequently we can find $x_1, x_2 \in X$, such that $Sx_1 = y$ and $Tx_2 = y$. This means that $S^{x_1} = T^{x_2} \neq 0$, that is, by Definition 11.8 the operators S and T are dependent. ■

11.13. Lemma. *Let X be a $C(K)$-module, $T \in dh_n(X)$ and $m \in \mathbf{N}$. Then*

1) $E_{T^m} \subseteq E_T$.
2) *For each $t \in E_{T^m}$ the points $\varphi_T(t), \varphi_T^2(t), \ldots, \varphi_T^{m-1}(t)$ lie in E_T.*
3) $\varphi_{T^m} \equiv \varphi_T^m$ *on* E_{T^m}

Proof. We omit a rather straightforward proof of this lemma.

In the next corollary X satisfies $(\diamond)$, $T \in dh_n(X)$, and

$$F_T^{(m)} = \{t \in E_T : (\forall k \leq m-1)(\varphi_T^k(t) \in E_T)\,\&\,(\exists k_0 \leq m-1)(\varphi_T^{k_0}(t) = t)\}$$

is the subset of E_T of all points of period $\leq m$, $m \in \mathbf{N}$, with respect to φ_T. Let also $V_T^{(m)} = \varphi_T^{-m}(F_T^{(m)})$.

11.14. Corollary. *Assume that $x = 0$ is the only element in X for which $supp(x) \subset V_T^{(k)}$ for some $k \in \mathbf{N}$. Then $T^m \,\delta\, T^n$ for all $n \neq m \in \mathbf{N} \cup \{0\}$.*

Proof. Indeed if $T^p \,\bar{\delta}\, T^q$ for some $p > q$, then from Proposition 11.12 and Lemma 11.13 it follows immediately that there exists $0 \neq x \in X$ satisfying $supp(x) \subset \varphi_T^{-q}(F_T^{p-q}) \subset \varphi_T^{-m}(F_T^m)$, where $m = \max(q, p-q)$; a contradiction. ■

11.15. Corollary. *Let T be a d-automorphism on a $C(K)$-module X satisfying $(\diamond)$, and let $0 \le m < n$. The following statements are equivalent.*

(1) $T^m \,\bar{\delta}\, T^n$.

(2) $I \,\bar{\delta}\, T^{n-m}$.

(3) *There is a non-zero $x \in X$ such that $T^{n-m} f x = f T^{n-m} x$ for each $f \in C(K)$.*

Proof. Since T is a d-automorphism, the mapping φ_T is a homeomorphism. The equivalence (1) $\Leftrightarrow$(2) follows from Corollary 11.14 and Proposition 11.12.

To prove (2) $\Leftrightarrow$(3) suppose first that (2) holds. Then by Proposition11.12 there exists a non-zero $x \in X$ with $supp(x) \subset \{s : \varphi_T^{n-m}(s) = s\}$. Let $p = n - m$. Then for each $f \in C(K)$ we have

$$T^{-p} f T^p x = (f \circ \varphi_T^{-p}) T^{-p} T^p x = (f \circ \varphi_T^{-p}) x = f x,$$

and consequently, multiplying by T^p, we obtain $f T^p x = T^p f x$, as desired.

Conversely, assume that $f T^p x = T^p f x$ for some $x \neq 0$ and all $f \in C(K)$. Then $[f \circ \varphi_T^p - f] T^p x = 0$, whence $\varphi_T^p(s) = s$ for any $s \in supp(T^p x)$. Therefore, by Proposition 11.12, $T^p \,\bar{\delta}\, I$. ■

11.16. Remark. The previous corollary remains true even without the assumption $(\diamond)$. Indeed, by Theorem 6.7 we can consider T as a d-automorphism of X viewed as a $Z(X)$-module, which is an operator module by Definition 6.1 and Remark 6.1.1. It remains to notice that by Definition 11.7, $T^m \delta T^n$ in $C(K)$-module X if and only if $T^m \delta T^n$ in $Z(X)$-module X.

11.17. Corollary. *Let S, T be two independent d-automorphisms of a $C(K)$-module X satisfying $(\diamond)$, and $A, B \in (C(K))^{com}$. Then $AS \,\bar{\delta}\, BT$ and $SA \,\bar{\delta}\, TB$.*

Proof. The equality $ASf = A(f \circ \varphi_S)S = (f \circ \varphi_S)AS$, where $f \in C(K)$, shows that $AS \in dh_n(X)$. Similarly $BT \in dh_n(X)$. We can easily verify that $\varphi_{AS}|E_{AS} = \varphi_S|E_{AS}$ and $\varphi_{BT}|E_{BT} = \varphi_T|E_{BT}$. If $AS \,\bar{\delta}\, BT$, then by Proposition 11.12 there would exist a non-zero $y \in X$ such that $supp(y) \subset \{q \in E_{AS} \cap E_{BT} : \varphi_S(q) = \varphi_T(q)\}$. Applying to this the second part of Proposition 11.14 (it is possible since S and T are onto) we conclude that $S \,\bar{\delta}\, T$, a contradiction. Therefore $AS \,\delta\, BT$.

Since $SA = SAS^{-1}S$ and $TB = TBT^{-1}T$ and since the operators SAS^{-1}, TBT^{-1} belong to $(C(K))^{com}$, the already proven part implies the remaining statement. ∎

11.18. Corollary. *Let T be a d-automorphism of a $C(K)$-module X satisfying ($\diamond$) and let $T^m \,\delta\, T^n$ for some $0 \le m < n$. Also let $A \in (C(K))^{com}$ and $T_1 := AT$. Then $T_1^m \,\delta\, T_1^n$.*

Proof. Let $A_k := A(TAT^{-1})(T^2AT^{-2})\dots(T^{k-1}AT^{-k+1})$. Clearly the operator A_k belongs to $(C(K))^{com}$. Now observe that $T_1^m = A_mT^m$ and $T_1^n = A_nT^n$. By the previous corollary, $T_1^m \,\delta\, T_1^n$. ∎

The Dedekind complete vector lattices have many advantages over arbitrary vector lattices. Much more is known about them. In the next definition we single out a class of $C(K)$-modules, which are their counterparts.

11.19. Definition. *An operator $C(K)$-module X is said to be order complete (resp. order σ-complete) provided the following two conditions are satisfied:*

(1) *The compact space K is extremally disconnected (resp. basically disconnected).*

(2) *For each $x \in X$ the set $\{f \in C(K) : fx = 0\}$ is a band in $C(K)$.*

Recall that condition (1) is equivalent to the order completeness (resp. order σ-completeness) of vector lattice $C(K)$.

11.19.1. Note that if K is extremally disconnected, then condition (2) above is equivalent to the following one: for each $x \in X$ its support set *supp*(x) is a clopen subset of K.

11.19.2. Remark. Let X be a Dedekind complete Banach lattice. Then, as we know, X is an operator $C(K)$-module, where $K = Q(X)$ is the Stone space of X. Obviously this $C(K)$-module is order complete. This explains why the concept of order complete $C(K)$-modules generalizes that of order complete vector lattices.

11.20. Proposition. ([Ral], Proposition 4.3.) *Let X be an operator $C(K)$-module, and let either of the following two conditions hold.*

(1) *X is order σ-complete and $X = X(x_0)$ for some $x_0 \in X$.*

(2) *X is order complete.*

Then $Z(X) = C(K)$.

Proof. Rall's paper cited above contains the proof for only the special case when X is a cyclic $C(K)$-module . The proof of the general case goes as follows (see also [HO1] for an independent proof). Take $H \in Z(X)$ and arbitrary elements x and y in X. By the above-mentioned Rall result there exist functions $f_x, f_y \in C(K)$ such that $H|X(x) = f_x|X(x)$ and $H|X(y) = f_y|X(y)$. We will show that $f_x \equiv f_y$ on the set $A = supp(x) \cap supp(y)$. Notice that we have equalities $Hx = f_x x$, $Hy = f_y y$, and $H(x+y) = f_{x+y}(x+y)$, where f_{x+y} exists by the same reason as above. As a result we get $f_x x + f_y y = f_{x+y}(x+y)$. Let B be any clopen subset of A on which $f_x \neq f_{x+y}$. Then obviously, $\chi_B x = \frac{f_{x+y}-f_y}{f_x - f_{x+y}} \chi_B y$. Applying to this equality the operator H again we obtain $f_x \chi_B x = f_y \frac{f_{x+y}-f_y}{f_x - f_{x+y}} \chi_B y$, which implies that $f_x \equiv f_y$ on B. Similarly, $f_x \equiv f_y$ on each clopen subset B, where $f_y \neq f_{x+y}$. The remaining possibility that $f_x = f_{x+y}$ and $f_y = f_{x+y}$ on some clopen subset B also implies that $f_x \equiv f_y$ on B. Therefore, we can conclude that $f_x \equiv f_y$ on A. Since K is extremally disconnected we can piece together all the functions f_x, i.e., there exists $f \in C(K)$ such that $fx = Hx$ for every $x \in X$. ∎

11.21. Remark. We can give a slightly more general definition of an order complete $C(K)$-module by postulating the following two conditions:

(1) *The maximal ideal space of the algebra $Z(X)$ is extremally disconnected.*

(2) *For each $x \in X$ the set $\{f \in Z(X) : fx = 0\}$ is a band in $Z(X)$.*

By virtue of Proposition 11.20 this definition is indeed more general than Definition 11.19. Its practicality depends on the following observation. If we consider $X = L^1[0,1]$ as a $C[0,1]$-module, then obviously this module is not order complete in the sense of Definition 11.19, though it is order complete in the more general sense described above. So, if one needs to incorporate this type of module into consideration, then the second definition would be appropriate. However, though less general, Definition 11.19 is more attractive and simple, since it uses only $C(K)$, rather than $Z(X)$. For our present purposes Definition 11.19 is quite satisfactory.

11.22. Corollary. *Let X be an order complete $C(K)$-module and let X_0 be an arbitrary closed $C(K)$-submodule of X. Let $K_0 = supp(X_0) := cl\{\bigcup_{x \in X_0} supp(x)\}$. Then $Z(X_0) = C(K_0)$, that is, $Z(X_0)$ may be identified with the subspace $\chi_{K_0}C(K)$ of $C(K)$. In particular, $Z(X(x)) = C(supp(x))$ for each $x \in X$.*

11.22.1. Let $\mathcal{A}$ be a subalgebra of $L(X)$, where X is a Banach space. We say that $\mathcal{A}$ is **strictly reflexive**, provided for each closed $\mathcal{A}$-invariant subspace V of X the subalgebra $\mathcal{A}|V = \{A|V : A \in \mathcal{A}\}$ of $L(V)$ is reflexive (see 6.2.3) and closed in the strong operator topology $\tau_s(V)$ on $L(V)$.

11.22.2. Corollary. *Let X be an operator $C(K)$-module. The following two statements are equivalent.*

(1) *X is order complete.*

(2) *$\mathcal{A} = C(K)$ is strictly reflexive.*

Proof. The implication (1) $\Rightarrow$(2) follows from Corollary 11.22 and Theorem 6.2. Let us prove the converse implication. Take an arbitrary open subset G of K and consider the subspace $V = cl\{x_1 + x_2 : x_1, x_2 \in X, supp(x_1) \subset G,\ supp(x_2) \subset int(K\backslash G)\}$ of X. Obviously V is an operator $C(K)$-module and the function $\chi_{cl(G)}$ belongs to $Z(V)$. Since $C(K)|V$ is $\tau_s(V)$-closed, by Theorem 6.2, $Z(V) = C(K)|V$, and therefore the function $\chi_{cl(G)} \in C(K)$, i.e., $cl(G)$ is a clopen subset. Hence K is extremally disconnected.

Take now an arbitrary $x \in X$, and consider the subspace $V = cl\{X(x) + \{v \in X : supp(v) \cap supp(x) = \emptyset\}\}$. Using again reflexivity of $C(K)|V$ we can conclude that $\chi_{supp(x)} \in C(K)$, i.e., $supp(x)$ is clopen. By 11.19.1 this is equivalent to condition (2) of Definition 11.19. This completes the proof. ∎

11.22.3. Corollary. *Let X be an order complete $C(K)$-module and $x_1, x_2 \in X$. Then*

$$x_1 \,\mathrm{d}\, x_2 \Longleftrightarrow supp(x_1) \cap supp(x_2) = \emptyset.$$

Proof. As we know, $x_1 \,\mathrm{d}\, x_2$ is equivalent to the disjointness of these elements in the Banach lattice $X(x_1 + x_2)$. On the other hand, the latter statement is equivalent to $supp_Q(x_1) \cap supp_Q(x_2) = \emptyset$, where Q is the Stone space of $X(x_1 + x_2)$. Since, by Corollary 11.22, Q is a clopen subset of K we are done. ■

11.22.4. Remark. In view of the previous corollary, Theorem 10.7 implies that $T^{**} \in dh(X^{**}, Y^{**})$ for each $T \in dh(X, Y)$

11.23. Proposition. *Let X be a $C(K_1)$-module, Y be an order complete $C(K_2)$-module and $T \in dh(X, Y)$. Then $T \in dh_n(X, Y)$, the set E_T, introduced in* 11.11, *coincides with $supp(TX)$, and consequently E_T is a clopen subset of K_2.*

Proof. The set $E_0 := supp(TX)$ is, by definition, the closure of the open set $\bigcup_{x \in X} supp(Tx)$, and therefore it is clopen since K_2 is extremally disconnected.

Let Y_0 be the closure of Y_{TX} in Y. Then by Corollary 11.22, Y_0 is an order complete $C(E_0)$-module and $Z(Y_0) = C(E_0)$. Theorem 6.7 implies then that $T \in dh_n(X, Y)$. The equality $E_T = supp(TX)$ is obvious. ■

11.24. Proposition. *Let X and Y be as in the previous Proposition, $S, T \in dh(X, Y)$ and let $\{q \in E_S \cap E_T : \varphi_S(q) = \varphi_T(q)\}$ be nowhere dense in K_2. Then $S \,\delta\, T$.*

Proof. The validity of this proposition follows immediately from Propositions 11.23 and 11.12, and the fact that $supp(y)$ is a clopen subset of K_2 for each $y \in Y$. ■

11.25. Remark. If X, Y are Dedekind complete Banach lattices and $S, T \in dh(X, Y)$, then Definition 11.2 shows that the condition of Proposition 11.24 is not only sufficient but also necessary for the independence of S and T.

11.26. Proposition. *Let X be an order complete $C(K)$-module and $T \in dh(X)$. The following conditions are equivalent.*

(1) *The powers of T are independent, i.e., $T^m \,\delta\, T^n$ for all $m \neq n \in \mathbf{N} \cup \{0\}$*

(2) *For arbitrary $m, n \in \mathbf{N} \cup \{0\}$, $m < n$, the set $V_T^{(m)}$ is nowhere dense in E_{T^n}.* (*Definition of $V_T^{(m)}$ precedes Corollary* 11.14.)

Proof. The implication (2) $\Rightarrow$ (1) is true by Proposition 11.24. To prove the converse implication assume, by way of contradiction, that (2) does not hold. Then for some $0 \le m < n$ the mappings φ_{T^m} and φ_{T^n} coincide on a non-empty clopen subset E of E_{T^n}. By the definition of the latter set (and since X is order complete) there exists an $x \in X$ such that $supp(T^n x) \cap E \neq \emptyset$. Take a non-zero $f \in Z(X)$ with $supp(f) \subset supp(T^n x) \cap E$ and let $x_1 = T^{n-m}x$. Then $(fT^n)^x = (fT^m)^{x_1} \neq 0$, which (by Definition 11.8) means that $T^n \,\bar{\delta}\, T^m$, a contradiction. ∎

One of the most important classes of operators between vector lattices is the class of order continuous operators. For example, we have seen in Part I some pleasant properties of order continuous d-homomorphisms on vector lattices. The next definition establishes a corresponding counterpart for $C(K)$-modules.

11.27. Definition. *Let $T \in dh(X,Y)$, where X is a $C(K_1)$-module and Y is a $C(K_2)$-module. T is said to be order continuous if for any $x \in X$ the restriction $T|X(x)$ is an order continuous operator from the vector lattice $X(x)$ to the vector lattice $Y(Tx)$.*

11.28. Remark. Let X and Y be order complete modules. Then $T \in dh(X,Y)$ is order continuous if and only if the mapping φ_T is open. The proof is immediate from Definition 11.27 and Corollary 11.22.

11.29. Corollary. *Let X be an order complete $C(K)$-module, and $T \in dh(X)$ be order continuous. Then $T^m \,\delta\, T^n$ for all $m \neq n \in \mathbf{N} \cup \{0\}$ if and only if $I \,\delta\, T^n$ for all $n \in \mathbf{N}$.*

Proof. Assume that $T^p \,\bar{\delta}\, T^q$ for some $p, q \in \mathbf{N}$, $0 \le p < q$. Then $\varphi_T^{-p}(F_T^{q-p})$ is a non-empty clopen set. Since the map φ_T is open, the set F_T^{q-p} is clopen. Consequently $\varphi_T^i(t) \equiv t$ on some non-empty clopen subset of F_T^{q-p} for some $i \in \{0, 1, \ldots, q-p\}$. Therefore, $I \,\bar{\delta}\, T^i$, a contradiction. ∎

11.30. Corollary. *Let T be a d-automorphism on an order complete $C(K)$-module X. The following two statements are equivalent.*

(1) $T^m \,\delta\, T^n$ *for all* $m \neq n \in \mathbf{N} \cup \{0\}$.

(2) *The mapping* φ_T *does not have periodic points.*

Proof. Recall that according to a known theorem of Z. Frolik the set of fixed points of any homeomorphism on any extremally disconnected compact space is clopen. (We will prove this result in Appendix A, Corollary A.7.) In view of Proposition 11.26 this implies the desired equivalence. ■

11.31. Proposition. *For an arbitrary* $C(K)$*-module* X *its conjugate module* X^* (*in the sense of Definition* 7.8) *is order complete.*

Proof. Recall that $Z(X^*) = L^\infty_X$ is Dedekind complete, and hence we only need to verify condition (2) of Definition 11.19. Assume, by way of contradiction, that this condition fails. Then by 11.19.1, for some $0 \neq x^* \in X^*$ the set $supp(x^*)$ is nowhere dense in Q_{X^*}. Then for some $x \in X$ the measure $x^* \square x$ is non-zero and has a nowhere dense support in Q_{X^*}. This contradicts the fact that $x^* \square x$ is an order continuous functional on L^∞_X. ■

11.32. Proposition. *Let* X *be an operator* $C(K)$*-module and for each* $x \in X$ *the Banach lattice* $X(x)$ *has the order continuous norm. Then module* X *is order complete.*

Proof. By Theorem 9.10, $Z^*(X) = Z(X)^* (= L^\infty_X)$, and consequently $Z(X)$ is itself isomorphic to L^∞_X. Now we can apply the same arguments as in the previous proposition. ■

11.33. From now on we will denote by j_X, or merely by j, the operator of the canonical embedding of X into X^{**}. Since we do not use the embedding operator $j : X \to X^{**}$ and the representation operator simultaneously, there will arise no ambiguity in exploiting the same letter j to denote either of these two operators. Let us agree to write simply jT instead of $j \circ T$.

11.34. Lemma. *If* $T \in dh(X)$, *where* X *is an operator* $C(K)$*-module, then* $jT \in dh(X, X^{**})$.

Proof. We begin by verifying that the following characterization of disjointness in the $C(K)$-module X holds:

$$x_1 \mathrm{d} x_2 \Longleftrightarrow \inf_{f \in C(K)} \{\|f x_1\| + \|(\mathbf{1}-f)x_2\|\} = 0.$$

Indeed, if $x_1 \,\mathrm{d}\, x_2$, then $x_1, x_2 \in X(x_1+x_2)$, and they are disjoint in this lattice. We can find $f_n \in C(K)$ such that $f_n(x_1+x_2) - x_2 \to 0$ as $n \to \infty$. However, $\|f_n x_1 + f_n x_2 - x_2\| \geq \|f_n x_1\|$ because $x_1 \,\mathrm{d}\, x_2$. Therefore, $f_n x_1 \to 0$ and $(\mathbf{1}-f_n)x_2 \to 0$.

Conversely, if $\inf\{\|f x_1\| + \|(\mathbf{1}-f)x_2\| : f \in C(K)\} = 0$, then there exist $f_n \in C(K)$ such that $\|f_n x_1\| + \|(\mathbf{1}-f_n)x_2\| \to 0$, and hence $(\mathbf{1}-f_n)x_1 + (\mathbf{1}-f_n)x_2 \to x_1$ and $f_n x_1 + f_n x_2 \to x_2$. Therefore, $x_1, x_2 \in X(x_1+x_2)$. If the element $y = |x_1| \wedge |x_2| \neq 0$, then we would have $\|f_n x_1\| + \|(\mathbf{1}-f_n)x_2\| \geq \|f_n y\| + \|(\mathbf{1}-f_n)y\| \geq \|y\|$, a contradiction.

Since $f \in C(K)$ implies that $f^{**} \in Z(X^{**})$, we have $x_1 d x_2 \Longrightarrow (jTx_1)\,\mathrm{d}\,(jTx_2)$. ∎

11.35. Recall that for $u \in X^{**}$ its support is the following set:

$$supp(u) = \{q \in Q : hu \neq 0 \ \ \forall h \in C(Q) = Z(X^{**}) \text{ with } h(q) \neq 0\}.$$

From 11.22 and 11.31 it follows that $supp(u)$ is clopen and that algebra $Z(X^{**}(u))$ is isomorphic to $\chi_{supp(u)} Z(X^{**})$ for each $u \in X^{**}$.

For each $x \in X$, we denote by K_x the Gelfand space of the algebra $Z(X(x))$ and by Q_x the set $supp(jx)$.

Lemma 11.34 and standard arguments involving Theorem 6.7 imply that for each $x \in X$, there exists a unique continuous surjective mapping $\eta_x : Q_x \longrightarrow K_x$ such that

$$jT(fx) = (f \circ \eta_x) \cdot jTx, \quad f \in Z(X(x)).$$

Let X be a $C(K)$-module, $T_1, T_2 \in dh(X)$ and $T = T_1 T_2$. A criterion of independence of operators T and T_1 is presented in Proposition 11.36 below.

By Theorems 10.6 and 10.7, there exists a clopen subset $E \subset Q$ and a continuous mapping $\varphi : E \longrightarrow Q$ such that $T^{**}H = (H \circ \varphi) \cdot T^{**}$, $H \in Z(X^{**})$. The corresponding objects for operator T_1 will be denoted by E_1 and φ_1. Obviously $E \subset E_1$. Using this notation we have the following proposition.

11.36. Proposition. *Let X be an operator $C(K)$-module $T_1, T_2 \in dh(X)$, $T = T_1T_2$. Let $V = \{q \in E : \varphi(q) = \varphi_1(q)\}$. The following conditions are equivalent.*

(1) $T \,\overline{\delta}\, T_1$.

(2) *There are $x \in X$ and $g \in Z(X_{Tx})$ such that $gTx \neq 0$ and $gTfx = gT_1fT_2x$ for all $f \in C(K)$.*

(3) *There is $x \in X$ such that $Tx \neq 0$ and int $\eta_{Tx}(V \cap Q_{Tx}) \neq \emptyset$.*

Proof. (2) $\Longrightarrow$ (1). Let x and g be as in (2). Letting $x_1 = T_2x$, we obtain $gT^x = gT_1^{x_1}$ and therefore $T \,\overline{\delta}\, T_1$.

(1) $\Longrightarrow$ (2). Let $T \,\overline{\delta}\, T_1$. Then, there exist $x_1, x_2 \in X$, $g \in Z(X_{Tx})$, $g_1 \in Z(X_{T_1x_1})$ such that $gTfx = g_1T_1fx_1$ for all $f \in C(K)$ and $gTx = g_1T_1x_1 \neq 0$.

Let $X_1 = X_{T_1X}$, i.e., X_1 is the ideal in X generated by T_1X. By Theorem 6.7, there exists a homomorphism $\gamma(= \gamma_T) : C(K) \to Z(X_1)$ such that $T_1fx' = \gamma(f)T_1x'$ for all $f \in C(K)$ and $x' \in X$. Fix for a moment f and x', above. Then, for an arbitrary $g' \in Z(X_{T_1x'})$, the following two elements $g'\gamma(f)T_1x'$ and $\gamma(f)g'T_1x'$ are well-defined and equal because the restriction of the operator $\gamma(f)$ to $X_{T_1x'}$ belongs to $Z(X_{T_1x'})$.

Therefore for each $f \in C(K)$ we have

$$gTfx = g_1T_1fx_1 = \gamma(f)g_1T_1x_1 = \gamma(f)gTx = \gamma(f)gT_1T_2x = gT_1fT_2x,$$

that is, the implication (1) $\Longrightarrow$ (2) is proved.

(2) $\Longrightarrow$ (3). Let $\eta = \eta_{Tx}$. The equality $gTfx = gT_1fT_2x$ implies that

$$(g \circ \eta) \cdot T^{**}f^{**}jx = (g \circ \eta) \cdot T_1^{**}f^{**}T_2^{**}jx.$$

It is easy to show that there exists an element $u \in Z(X^*)$ such that

$$u^*T^{**}f^{**}jx = (g \circ \eta)T^{**}f^{**}jx \tag{$*$}$$

and

$$u^*T_1^{**}f^{**}T_2^{**}jx = (g \circ \eta) \cdot T_1^{**}f^{**}T_2^{**}Tx \tag{$**$}$$

(we can take for u any limit point of the sequence $\{g_n^*\}$, where $\{g_n\}$ is an arbitrary sequence in $C(K)$ satisfying $\|g_n\| \leq \|g\|$ and $g_nTx \longrightarrow gTx$ and existing by Proposition 6.4).

Theorem 9.15 clearly implies that for each $h \in Z(X^{**})$ there exists a net $\{f_\alpha\} \subset C(K)$ such that $\|f_\alpha\| \leq \|h\|$ and $f_\alpha^{**} \longrightarrow h$ in the topology $\sigma(L(X^{**}), X^{**} \otimes X^*)$. Consequently $(*)$ and $(**)$ imply that for each $h \in Z(X^{**})$

$$(g \circ \eta) \cdot T^{**}hjx = (g \circ \eta)T_1^{**}hT_2^{**}jx.$$

Therefore

$$(g \circ \eta)(h \circ \varphi - h \circ \varphi_1)T^{**}jx = 0.$$

Since h is arbitrary we can conclude that $\varphi \equiv \varphi_1$ on $supp((g \circ \eta)T^{**}jx)$. This proves (2) $\Longrightarrow$ (3).

(3) $\Longrightarrow$ (2). Let $x \in X$ satisfies (3). Pick $g \in Z(X_{Tx})$ such that $g \neq 0$ and $supp(g) \subset \eta_{Tx}(V \cap Q_{Tx})$. The operators $T : X(x) \to X(Tx)$ and $T_1 : X(T_2x) \to X(T(x)$ generate continuous mappings:

$$\psi : K_{Tx} \longrightarrow K_x \quad \text{and} \quad \psi_1 : K_{Tx} \longrightarrow K_{T_2x},$$

such that $Tfx = (f \circ \psi) \cdot Tx$ and $T_1fT_2x = (f \circ \psi_1) \cdot Tx$ for all $f \in C(K)$. Therefore the following two equalities hold:

$$jTfx = (f \circ \psi \circ \eta) \cdot jTx = (f^{**} \circ \varphi) \cdot jTx$$
$$jT_1fT_2x = (f \circ \varphi_1 \circ \eta) \cdot jTx = (f^{**} \circ \varphi_1) \cdot jTx,$$

where we consider operator f^{**} as an element of $Z(X^{**}) = C(Q)$.

Let $q \in Q_{Tx}$ and $g(\eta(q)) \neq 0$. Then, there exists $p \in V \cap Q_{Tx}$ such that $\eta(p) = \eta(q)$. Consequently,

$$\begin{aligned}(g \circ \eta) \cdot (f^{**} \circ \varphi - f^{**} \circ \varphi_1)(q) =\\ = (g \circ \eta) \cdot (f \circ \psi \circ \eta - f \circ \psi_1 \circ \eta)(q) =\\ = (g \circ \eta) \cdot (f \circ \psi \circ \eta - f \circ \psi_1 \circ \eta)(p) =\\ = (g \circ \eta) \cdot (f^{**} \circ \varphi - f^{**} \circ \varphi_1)(p) = 0.\end{aligned}$$

This gives $(g \circ \eta) \cdot (f^{**} \circ \varphi - f^{**} \circ \varphi_1) j T x = 0$, i.e., $gTfx = gT_1 f T_2 x$. Since f is arbitrary, this proves (3) $\Longrightarrow$ (2). ■

11.37. Remark. An inspection of the proof of the last implication reveals that in actuality the following statement is correct. *If $T \in dh(X)$, where X is an order complete $C(K)$-module, and $\varphi_{T^m} = \varphi_{T^n}$ on a clopen subset F of K, then $\varphi_{(T^{**})^m} = \varphi_{(T^{**})^n}$ on the set $\eta^{-1}(F)$, where η is a continuous mapping of $Q(X^{**})$ onto K, corresponding to the natural imbedding of $C(K)$ into $Z(X^{**})$.*

PART III. Spectral properties

12. Spectrum of d-endomorphisms of $C(K)$-modules

The main result of this section (Theorem 12.11) asserts that if the powers of a d-endomorphism T on a $C(K)$-module X are independent and X satisfies a very mild additional assumption, then the spectrum $\sigma(T)$ of T is rotation invariant. Imposing some additional assumptions on T and X, we will be able to obtain more detailed information about $\sigma(T)$. In Section 13 we will deal with d-endomorphisms on Banach lattices and for their spectrum, we obtain stronger results, which are broad generalizations of the results on this subject contained in [AAK], [AH], [AL], [HP], [Kit1], [Lat], [Mon] and [Pet].

Now we are going to introduce some notation which will be used throughout this and the next sections.

12.1. Symbol $\sigma(T)$ denotes, as usual, the spectrum of a continuous operator T. We say that $\sigma(T)$ is rotation invariant provided $e^{i\theta}\sigma(T) = \sigma(T)$ for all $\theta \in [0, 2\pi]$. Recall the definitions of the following two complimentary subsets of $\sigma(T)$. The approximate point spectrum $\sigma_a(T) = \{\lambda \in \sigma(T) : \exists \{x_n\} \subset X, \|x_n\| = 1,\ Tx_n - \lambda x_n \to 0\}$ and the residual spectrum $\sigma_r(T) = \sigma(T) \backslash \sigma_a(T)$. It is well known that $\sigma_r(T)$ consists of all those points $\lambda \in \sigma(T)$ for which the operator $T - \lambda I$ has a left inverse (on its image) and that $\sigma_r(T)$ is open in $\mathbf{C}$. This immediately implies that $\sigma(T)$ is rotation invariant provided $\sigma_a(T)$ is.

12.2. For simplicity of formulations we agree that the words "disc" and "annulus" always mean a closed disc and a closed annulus, respectively, centered at zero. The set $\{0\}$ is considered to be a disc of zero radius and this allows us to not single out the case $\sigma(T) = \{0\}$.

12.3. If Y is a subspace of X and $TY \subset Y$, then the expression that $\sigma(T|Y)$ is a disc or an annulus means that $\sigma(T|Y)$ is a disc or annulus provided $Y \neq \{0\}$, or $\sigma(T|Y) = \emptyset$ if $Y = \{0\}$.

12.4. The spectral radius of an operator T is denoted by $r(T)$. We denote by Γ the unit circle, and by D_r the (closed) disc of radius r centered at 0.

12.5. Let K be a compact Hausdorff space, E a closed subset of K and φ a continuous map from E into K. We say that a point $t \in E$ has period $m \in \mathbf{N}$ with respect to φ and write $p_\varphi(t) = m$ if $\varphi(t) \in E$, $\varphi^2(t) = \varphi(\varphi(t)) \in E, \ldots, \varphi^{m-1}(t) \in E$, $\varphi^m(t) = t$ and $\varphi^r(t) \neq t$ for $r < m$. By symbol $F^{(m)}$ (or $F_\varphi^{(m)}$ if we want to specify the generating mapping explicitly) we denote the subset of E of all the points with period less than or equal to m, that is, $F^{(m)} = \{t \in E : p_\varphi(t) \leq m\}$.

The inverse image of order m of the set $F^{(m)}$, that is, the set $\varphi^{-m}(F^{(m)})$ will be denoted by $V^{(m)}$.

In the case when K is extremally disconnected, we will also need the following clopen subsets of K:

$$G^{(m)} = int\{t \in E : p_\varphi(t) = m\}$$

and

$$H^{(m)} = cl \bigcup_{k=0}^{\infty} \varphi^{-k}(G^{(m)}).$$

The characteristic functions of these two sets will be denoted by $g^{(m)}$ and $h^{(m)}$, respectively.

12.6. Recall [KA, page 373] that an extremally disconnected Hausdorff compact space K is called **hyperstonean** if there exists a regular Borel measure μ on K such that $\mu(A) = 0$ if and only if A is nowhere dense in K. In this case the space $C(K)$ is isometrically and order isomorphic to the space $L^\infty(K, \mu)$.

We will need the following three statements proved in [Kit2]. In each of them K is extremally disconnected, E is a clopen subset of K and $\varphi : E \to K$ is a continuous open mapping. The proofs of these lemmas are presented in Appendices A and B.

12.6.1. Lemma. *Assume that* $F^{(m)} = \emptyset$. *Take an arbitrary* $\alpha \in \mathbf{C}$ *with* $|\alpha| = 1$. *Then, there exists an* $f \in C(K)$ *such that* $|f| \equiv 1$ *and* $|f(\varphi(t)) - \alpha f(t)| < \frac{2\pi}{m+1}$ *for all* $t \in E$.

12.6.2. Lemma. *If* K *is hyperstonean, then the set* $F^{(m)}$ *is clopen in* K *for each* $m \in \mathbf{N}$.

12.6.3. Lemma. *Let* K *be hyperstonean and* $\alpha \in \Gamma$. *Then for each* $m \in \mathbf{N}$ *there exists a function* $f \in C(K)$ *such that* $|f| \equiv 1$ *on* $K \backslash V^{(m)}$ *and* $|f(\varphi(t)) - \alpha f(t)| \leq (2\pi + 1)/m$ *for all* $t \in E$.

12.7. Let X be a $C(K)$-module and $x \in X$. Recall that $supp(x)$, or $supp_K(x)$ if there is a danger of ambiguity – the support set of x – is the smallest closed subset E of K such that $fx = 0$ whenever $f \in C(K)$ does not "live" on E, that is, $f(t) = 0$ for all $t \in E$. If X is order complete (Definition 11.19), then $supp(x)$ is a clopen subset of K for each $x \in X$.

12.8. As in 11.35, we denote by Q the Gelfand (= Stone) space of the algebra $Z(X^{**})$, by K_x the Gelfand space of the algebra $Z(X(x))$ and by η_x the continuous mapping from $Q_x = supp_Q(jx)$ onto K_x. Also we introduce a continuous mapping $\eta : Q \longrightarrow K$ which is generated by the canonical embedding of $C(K)$ into $Z(X^{**}) = C(Q)$.

12.9. Let $T \in dh_n(X)$, where X is a $C(K)$-module. In 11.11.1 we have considered the ideal $J_T = \{f \in C(K) : fT = 0\}$ in $C(K)$ and have noticed that T generates in a canonical way a homomorphism γ_T of the algebra $C(K)$ into the quotient algebra $C(K)/J_T$. In its turn this homomorphism γ_T generates a continuous mapping φ_T of the zero set E_T of the ideal J_T into K. The sets $F_{\varphi_T}^{(m)}$ and $V_{\varphi_T}^{(m)}$ generated by the mapping φ_T (in accordance with 12.5) will be denoted by $F_T^{(m)}$ and $V_T^{(m)}$, respectively. If E_T is extremally disconnected,then for this mapping φ_T, we can also consider the sets $G_T^{(m)}, H_T^{(m)}$, and their characteristic functions $g_T^{(m)}$, $h_T^{(m)}$. Occasionally, when it does not cause any ambiguity, we will omit index T in the above notations.

Whenever we deal with the second conjugate operator T^{**}, which, as we know, belongs to $dh(X^{**})$, the above defined set $E_{T^{**}}$ and mapping $\varphi_{T^{**}}$ are nothing else but the corresponding objects described in Theorem 10.7.

If X is an order complete $C(K)$-module, then (by Proposition 11.23) each d-endomorphism $T \in dh(X)$ belongs to $dh_n(X)$.

12.10. Recall [KA] that the norm in a Banach lattice X satisfies condition (C) (= is weakly Fatou = is semi-continuous) if $(0 \leq x_\alpha \uparrow x)$ implies $(\|x_\alpha\| \uparrow \|x\|)$.

12.10.1. Definition. *We say that a $C(K)$-module X satisfies condition* (C) *if each Banach lattice $X(x)$, $x \in X$, satisfies* (C).

Now, at last, we can formulate our first result.

12.11. Theorem. *Let X be a $C(K)$-module satisfying* (C) *and $T \in dh(X)$. If $T^p \ \delta \ T^q$ for each distinct $p, q \in \mathbf{N} \cup \{0\}$, then the spectrum $\sigma(T)$ is rotation invariant.*

Proof. It is enough to show that if $1 \in \sigma_a(T)$, then $\Gamma \subset \sigma_a(T)$. Fix a number $\alpha \in \Gamma$ and let $x_n \in X, \|x_n\| = 1$ satisfy $Tx_n - x_n \longrightarrow 0$.

As we know the compact space Q is hyperstonean and the mapping $\varphi_{T^{**}}$ is open (Theorem 10.7). Therefore, for each $m \in \mathbf{N}$ we can find a function $f \in C(Q)$ satisfying the conclusions of Lemma 12.6.3.

Let $p := m! + m$. Let $y_n := T^p x_n$. The mappings $\varphi^p_{T^{**}}$ and $\varphi^m_{T^{**}}$ coincide on the set $V^{(m)}_{T^{**}}$. Since $T^m \ \delta \ T^p$, Proposition 11.36 implies that the set $K_n = \eta_{y_n}(V^{(m)}_{T^{**}} \cap supp(jy_n))$ is nowhere dense in K_{y_n} for each n.

Obviously, $\|y_n\| \longrightarrow 1$ and $Ty_n - y_n \longrightarrow 0$. Letting $z_n := fjy_n$, we are going to show that

$$\liminf \|T^{**}z_n - \alpha z_n\| \leq \frac{2\pi + 1}{m} \|T\| \tag{12.1}$$

and

$$\lim_n \|z_n\| = 1. \tag{12.2}$$

Since m is arbitrary, (12.1) and (12.2) imply that $\alpha \in \sigma_a(T^{**})$ and this proves the theorem (keep in mind that $\sigma_a(T) = \sigma_a(T^{**})$).

To prove (12.1) recall (see Theorem 10.6) that by the definition of the set $E = E_{T^{**}}$ we have $T^{**}(jy_n) = \chi_E T^{**}(jy_n)$. Since $T^{**}(jy_n) - jy_n \longrightarrow 0$, the previous equality implies that $(\mathbf{1}-\chi_E)jy_n \longrightarrow 0$, and consequently $(\mathbf{1}-\chi_E)z_n = (\mathbf{1}-\chi_E)fjy_n \to 0$. Further we have

$$\begin{aligned}&T^{**}z_n - \alpha z_n\\ &= -\alpha(\mathbf{1}-\chi_E)z_n + T^{**}z_n - \alpha\chi_E z_n\\ &= -\alpha(\mathbf{1}-\chi_E)z_n + \gamma_{T^{**}}(f)T^{**}jy_n - \alpha\chi_E fjy_n\\ &= -\alpha(\mathbf{1}-\chi_E)z_n + (\gamma_{T^{**}}(f) - \alpha\chi_E f)T^{**}jy_n + \alpha\chi_E f(T^{**}jy_n - jy_n),\end{aligned}$$

whence $\liminf \|T^{**}z_n - \alpha z_n\| = \liminf \|(\gamma_{T^{**}}(f) - \alpha\chi_E f)T^{**}jy_n\|$. By our choice of f we have $\|\gamma_{T^{**}}(f) - \alpha\chi_E f\|_\infty \le \frac{2\pi+1}{m}$ and (12.1) is proved.

It remains to prove (12.2). Since the set K_n is nowhere dense in K_{y_n}, we can find a net $\{y_\beta\} \subset X(y_n)$ such that $|y_\beta| \uparrow_\beta |y_n|$ and also $supp(y_\beta) \cap K_n = \emptyset$ for each β. Therefore, $supp_Q(jy_\beta) \cap \eta_{y_n}^{-1}(K_n) = \emptyset$ and note that the latter set contains the set $V_{T^{**}}^{(m)} \cap supp(jy_n)$. Since $|f| \equiv 1$ on $Q\backslash V_{T^{**}}^{(m)}$, we get $|jy_\beta| \le |fjy_n| = |z_n|$. But on the Banach lattice $X(y_n)$ the norm satisfies (C) and therefore $\lim_\beta \|y_\beta\| = \|y_n\|$. Consequently $\|z_n\| = \|y_n\|$. ■

12.12. Corollary. *Let X be a $C(K)$-module with condition* (C) *and let $T \in dh_n(X)$. Assume also that for each $m \in \mathbf{N}$ the following implication holds:*

$$supp(x) \subset V_T^{(m)} \Longrightarrow x = 0.$$

Then $\sigma(T)$ is rotation invariant.

Proof. By Corollary 11.14 the powers of T are pairwise independent. It remains to apply the previous theorem. ■

12.12.1. Remark. If we assume that for each $m \in \mathbf{N}$ the set $F_T^{(m)}$ is empty (in other words, φ_T has no periodic points), then the conclusion of the previous corollary remains valid without the assumption that the norm on X satisfies (C).

Indeed, in this case for each m the set $F_{T^{**}}^{(m)} = \emptyset$, and in view of Lemma 12.6.1 the proof of Theorem 12.11 may be carried through.

12.13. Corollary. *Let X be an order complete $C(K)$-module with condition* (C), $T \in dh(X)$ *and assume that for each $m \in \mathbf{N}$ the set $V_T^{(m)}$ is nowhere dense in K. Then $\sigma(T)$ is rotation invariant.*

Proof. By Proposition 11.24 the powers of T are pairwise independent, and so Theorem 12.11 is applicable. ■

Actually, the following much stronger result is true.

12.14. Proposition. *Let X be an order complete $C(K)$-module with condition* (C) *and $T \in dh(X)$. If for each $m \in \mathbf{N}$*

$$F_T^{(m)} \cap intV_T^{(m)} = \emptyset, \qquad (*)$$

then for each $T_1 \in Z(X)^{com}$ the spectrum $\sigma(T_1T)$, in particular $\sigma(T)$, is rotation invariant. Moreover, the condition $(*)$ is necessary in the following sense. If X is a Dedekind complete Banach lattice and $(*)$ does not hold, then there exists an operator $T_1 \in Z(X)$ for which $\sigma(T_1T)$ is not rotation invariant.

We have to postpone the proof of this proposition (until after 12.19) since it requires some additional information about $\sigma(T^{**})$. In the next remark we pose an interesting open question.

12.14.1. Remark. Let K be an extremally disconnected compact space and φ be a continuous mapping of K into itself. We do not know whether the following two implications are true.

(1) $\{\forall m : F_T^{(m)} \cap intV_T^{(m)} = \emptyset\} \Longrightarrow \{\forall m : intV_T^{(m)} = \emptyset\}$.

(2) $\{\forall m : intV_T^{(m)} = \emptyset\} \Longrightarrow \{\forall m : F_T^{(m)} = \emptyset\}$.

Both of these implications are true if the mapping φ is injective (see Corollary A.7 (Frolik)), or if φ is open and K is hyperstonean ([Kit2] and Lemma 12.6.2). If they are true without any additional assumptions, then Proposition 12.14 is true without assumption (C) of the semi-continuity of norm in X.

Now we turn our attention to the case where X is an order complete $C(K)$-module and T is an order continuous d-homomorphism. Recall that by Remark 11.28 the mapping $\varphi = \varphi_T : E_T \to K$ is open.

Let us introduce the following two spaces $X_m := h^{(m)}X$, $X^{(m)} := g^{(m)}X$ (see 12.5, where the function $g^{(m)}$ and $h^{(m)}$ are introduced), two following operators $T_m := T|X_m$, $T^{(m)} := g^{(m)}T|X^{(m)}$, and finally the following spectra $\sigma_m := \sigma(T_m)$ and $\sigma^{(m)} := \sigma(T^{(m)})$.

A comment explaining why T_m maps X_m into X_m is due. The definition of the set $H^{(m)}$ implies that $\varphi_T^{-1}(H^{(m)}) = H^{(m)}$, and thus $h^{(m)} \circ \varphi = h^{(m)}$. Therefore, for each $x \in X_m$ we have

$$T_m x = Tx = Th^{(m)}x = (h^{(m)} \circ \varphi_T)Tx = h^{(m)}Tx \in X_m.$$

In the terms of the introduced above sets $\sigma^{(m)}$ and σ_m we are able to analyze the rotation structure of the spectrum of d-homomorphisms. Our next result was motivated by a theorem of E. Scheffold (see Proposition 12.19.1, or [Sch], Theorem V.4.4) stating that the spectrum of each lattice homomorphism on a Banach lattice is cyclic. The next Proposition 12.15 presents a generalization of this result for arbitrary order continuous d-homomorphisms on $C(K)$-modules, while Proposition 12.18 improves Proposition 12.15 if we consider Banach lattices instead of $C(K)$-modules.

12.15. Proposition. *Let X be an order complete $C(K)$-module with* (C), *$T \in dh(X)$ and T be order continuous. Then*

(1) *$\sigma^{(m)} = \zeta_m \sigma^{(m)}$ and $\sigma_m = \zeta_m \sigma_m$, where $\zeta_m = \exp(\frac{2\pi i}{m})$.*

(2) *$\sigma_m(T) \subseteq \sigma(T)$ and $\sigma(T)\setminus \cup_m \sigma_m$ is rotation invariant.*

(3) *$\sigma_a(T_m)\setminus\sigma_a(T^{(m)}))$ is rotation invariant for each $m \in \mathbf{N}$.*

Proof. Since the set $G^{(m)}$ is extremally disconnected and φ_T is an open mapping on $G^{(m)}$, there exists a maximal clopen subset R such that $\varphi^k(R) \cap \varphi^l(R) = \emptyset$ for $0 \le k < l \le m-1$. This implies that $G^{(m)} = \cup_{k=1}^{m-1} \varphi^k(R)$ and $\varphi^m(R) = R$.

Let us introduce further a function $g(t)$ by letting $g(t) = \zeta_m^k$ for $t \in \varphi_T^k(R)$, $0 \le k \le m-1$, and $g(t) = \zeta_m^{-(l+1)} g(\varphi_T^{l+1}(t))$ for $t \in \varphi_T^{-(l+1)}(G^{(m)})\setminus\varphi_T^{-l}(G^{(m)})$ and $l = 0, 1, 2, \ldots$. This defines a continuous g on an open dense subset of $H^{(m)}$ and since $H^{(m)}$ is extremally disconnected, there exists a continuous extension of g to all of $H^{(m)}$. We retain the same letter g to denote this extension. In view of the definition $g(\varphi_T(t)) = \zeta_m g(t)$ for all $t \in H^{(m)}$. Finally, let $g \equiv 1$ on $K \backslash H^{(m)}$. We claim that $g^{-1}T_m g = \zeta_m T_m$ and $g^{-1}T^{(m)}g = \zeta_m T^{(m)}$, and this implies (1). We will verify the first equality only. To this end notice that $g \circ \varphi_T = \zeta_m g$. Now we have $g^{-1}T_m g = g^{-1}(g \circ \varphi_T)T_m = g^{-1}\zeta_m g T_m = \zeta_m T_m$ as claimed. The similarity of the operators T_m and $\zeta_m T_m$ implies that $\sigma_m = \zeta_m \sigma_m$.

(2) The mapping φ_T is open, hence $\varphi_T^{-1}(H^{(m)}) = H^{(m)}$, and therefore, as was shown, $h^{(m)}T = Th^{(m)}$. This implies that $\sigma_m \subset \sigma(T)$ in view of the following well-known fact: if an operator T on a Banach space E commutes with a projection onto a subspace E_0 of E, then $\sigma(T|E_0) \subseteq \sigma(T)$.

Let $0 \neq \lambda \in \sigma(T) \backslash \cup_m \sigma_m$ and let h_∞ be the characteristic function of the set $H_\infty := cl \cup_m H^{(m)}$. Clearly, $\varphi_T^{-1}(H_\infty) = H_\infty$, and therefore $Th_\infty = h_\infty T$. Now, the commutativity of the operator T and the projection h_∞ implies that $\sigma(T) \cup \{0\} = \sigma(h_\infty T) \cup \sigma((\mathbf{1} - h_\infty)T)$, and consequently only the following two cases are possible.

(a) $\lambda \in \sigma(h_\infty T)$. Without loss of generality we can assume that $\lambda \in \sigma_a(h_\infty T)$. Let $\{x_n\} \subset X$ be such that $\|x_n\| = 1$ and $h_\infty T x_n - \lambda x_n \to 0$. Since $\lambda \notin \cup_m \sigma_m$ we can conclude that $h^{(m)}x_n \to 0$ for each m. Indeed, $h_\infty T x_n - \lambda x_n \to 0$ clearly implies that $h^{(m)}h_\infty T x_n - \lambda h^{(m)}x_n \to 0$. Since $h_\infty h^{(m)} = h^{(m)}$ we actually have $h^{(m)}Tx_n - \lambda h^{(m)}x_n \to 0$, or in view of the commutativity, $Th^{(m)}x_n - \lambda h^{(m)}x_n \to 0$. But $\lambda \notin \sigma_m$ and this means that $h^{(m)}x_n \to 0$.

Let us fix an m and let $y_n := x_n - \sum_{k=1}^{m} h^{(k)}x_n$. Then $\|y_n\| \to 1$ and $h_\infty T y_n - y_n \to 0$ as $n \to \infty$. Let $\alpha \in \mathbf{C}$, $|\alpha| = 1$. By Lemma 12.6.1, there exists a function g continuous and unimodular on the set $\mathrm{cl}(\cup_{l=m+1}^{\infty} H^{(l)})$ and such that $|g \circ \varphi_T - \alpha g| \leq \frac{2\pi}{m}$ on this set. We extend g to the rest of K as zero.

Letting $z_n := gy_n$ we have $\|z_n\| = \|y_n\|$ and $\liminf \|Tz_n - \alpha z_n\| \leq 2\pi \frac{\|T\|}{m}$. Hence $\alpha \in \sigma_a(T)$.

(b) $\lambda \in \sigma((\mathbf{1} - h_\infty)T)$. Then $\lambda \in \sigma(T|(\mathbf{1} - h_\infty)X)$ because $\lambda \neq 0$ and $h_\infty T = Th_\infty$. But the operator T on the module $(\mathbf{1} - h_\infty)X$ satisfies all the conditions of Corollary 12.13, and therefore $\lambda\Gamma \subset \sigma(T|(\mathbf{1} - h_\infty)X) \subset \sigma(T) \cup \{0\}$.

(3) Let $\lambda \in \sigma_a(T_m) \backslash \sigma_a(T^{(m)})$. Pick $x_n \in X$ such that $\|x_n\| = 1$, $x_n = h^{(m)}x_n$

and $T_m x_n - \lambda x_n \to 0$ as $n \to \infty$. We claim that $g^{(m)} x_n \to 0$ as $n \to \infty$. To verify this we first recall that $\varphi_T(G^{(m)}) = G^{(m)}$, and thus $\varphi_T^{-1}(G^{(m)}) \supseteq G^{(m)}$. Therefore $g^{(m)} \cdot (g^{(m)} \circ \varphi_T) = g^{(m)}$.

As said above, $T_m x_n - \lambda x_n \to 0$, and hence $g^{(m)} T_m x_n - \lambda g^{(m)} x_n \to 0$. Now notice that $g^{(m)} T g^{(m)} = g^{(m)}(g^{(m)} \circ \varphi_T) T = g^{(m)} T$, and therefore we have $g^{(m)} T_m g^{(m)} x_n - \lambda g^{(m)} x_n \to 0$. This means that either $g^{(m)} x_n \to 0$ as $n \to \infty$ or else $\lambda \in \sigma_a(T_m)$. In view of our assumptions only the former case is possible.

To conclude the proof, fix an arbitrary $\alpha \in \Gamma$ and define a unimodular function g on $H^{(m)}$ in the following way: let $g \equiv 1$ on $G^{(m)}$ and let $g \equiv \alpha^{-k}$ on the set $\varphi_T^{-k}(G^{(m)}) \backslash \varphi_T^{-(k-1)}(G^{(m)})$ for $k \in \mathbf{N}$. After that we can extend g to $H^{(m)}$ by continuity.

The proof will be done as soon as we show that $T_m g x_n - \alpha \lambda g x_n \to 0$ as $n \to \infty$. To this end let us note that $g(\varphi_T(t)) = \alpha g(t)$ for each $t \in H^{(m)} \backslash G^{(m)}$. Consider now elements $y_n = x_n - g^{(m)} x_n$. For each m we have $supp(T_m y_n) \subseteq H^{(m)} \backslash G^{(m)}$, $x_n - y_n = g^{(m)} x_n \to 0$, and consequently $T_m y_n - \lambda y_n \to 0$ as $n \to \infty$. It remains to notice that $T_m g y_n - \alpha \lambda g y_n = (g \circ \varphi_T) T_m y_n - \alpha \lambda g y_n = \alpha g T_m y_n - \alpha \lambda g y_n = \alpha g (T_m y_n - \lambda y_n) \to 0$. ■

12.16. Remark. In the important special case of hyperstonean K, the previous Proposition 12.15 remains true even without the assumption (C). Indeed, in our proof we used (C) only once (in part 2(b)) when we claimed that $\sigma((\mathbf{1} - h_\infty)T)$ is rotation invariant. But if K is hyperstonean, then, by Lemma 12.6.2, the restriction of φ_T to the set $K \backslash \operatorname{cl}(\cup_m H^{(m)})$ does not have periodic points at all and we can use Remark 12.12.1 or Lemma 12.6.1. An application of this observation will be given in Example 14.25.

As we know (Theorems 10.6 and 10.7) the second conjugate T^{**} is order continuous and belongs to $dh(X^{**})$ for any $T \in dh(X)$. Therefore Proposition 12.15 is applicable to the operator T^{**}. However, it is not very surprising since we can show that module X^{**} always satisfies (C).

12.17. Remark. The inclusion $\sigma^{(m)} \subset \sigma(T) \cup \{0\}$ is not true in general. For example, take for X the two-sided space ℓ^2, and introduce on X the following structure of an ℓ^∞-module: if $f = (f_0, f_1, \ldots,) \in \ell^\infty$ and $x = \{x_n : n \in \mathbf{Z}\} \in X$, then

$$fx := (\ldots, f_0x_{-2}, f_0x_{-1}, f_0x_0, f_1x_1, f_2x_2, f_3x_3, \ldots).$$

Take the right shift operator for T. It is easy to verify that $\sigma(T) = \Gamma$, but $\sigma^{(1)} = D_1$. Notice also that though T is an invertible d-endomorphism, it easy to verify that T^{-1} is not a d-endomorphism. That is, we have got an example promised in 4.5.1.

However, if X is a Dedekind complete Banach lattice (considered with its standard $Z(X)$-module structure), then the inclusion $\sigma^{(m)} \subset \sigma(T) \cup \{0\}$ holds and, moreover, we can considerably improve Proposition 12.15. To do so we need the next well-known fact.

12.18.0. Let $T : X \to X$ be a continuous operator on a Banach space, let X_0 be a T-invariant subspace of X, and let $\overset{\bullet}{T}$ be the quotient operator generated by T on $X/_{X_0}$. Consider the three operators $T, T|X_0$ and $\overset{\bullet}{T}$.

(1) If any two of the above operators are invertible, then the third is invertible, too.

(2) If T is invertible and $T|X_0$ is one-to-one but not onto, then $\overset{\bullet}{T}$ is onto but not one-to-one.

12.18. Proposition. *Let X be a Dedekind complete Banach lattice with condition (C) and let $T \in dh(X)$ be order continuous. Then $\sigma^{(m)} \subset \sigma(T)$ and $\sigma(T) \setminus \cup_m \sigma^{(m)}$ is rotation invariant.*

Proof. To prove that $\sigma^{(m)} \subseteq \sigma(T)$ take an arbitrary $\lambda \notin \sigma(T)$ and then verify that $\lambda \notin \sigma^{(m)}$. Notice that $(T - \lambda I)X = X$. Let $x = g^{(m)}x \in X^{(m)}(= g^{(m)}X)$. Then there exists $y \in X$ such that $Ty - \lambda y = g^{(m)}x$, and consequently $g^{(m)}Ty - \lambda g^{(m)}y = g^{(m)}x$. But $g^{(m)}Tg^{(m)}y = g^{(m)}(g^{(m)} \circ \varphi_T)Ty = g^{(m)}Ty$, whence $g^{(m)}Tg^{(m)}y - \lambda g^{(m)}y = g^{(m)}x$ and $(T^{(m)} - \lambda I)X^{(m)} = X^{(m)}$. Now we have only to show that the operator $T^{(m)} - \lambda I$ is injective on $X^{(m)}$. To this end, notice that the proof of Proposition 12.15(1) shows that for each $k = 1, 2, \ldots, m-1$, the operator $T^{(m)}$ is similar to $\zeta_m^k T^{(m)}$, where $\zeta_m = \exp(2\pi i/m)$. Consequently we have $(T^{(m)} - \lambda\zeta_m^k I)X^{(m)} = X^{(m)}$ for each $k = 1, 2, \ldots, m-1$, and hence $((T^{(m)})^m - \lambda^m I)X^{(m)} = X^{(m)}$. Since the operator

$(T^{(m)})^m$ commutes with $Z(X^{(m)})$ (because $G^{(m)}$ consists of points of period m) and since $X^{(m)}$ is a Dedekind complete Banach lattice, we can conclude that $(T^{(m)})^m$ belongs to $Z(X^{(m)})$ and therefore can be represented as a multiplication operator by a continuous function. It is trivial to see that such an operator is onto if and only if it is invertible. Hence $ker(T^{(m)} - \lambda I) = 0$ and $\lambda \notin \sigma^{(m)}$.

To show that $\sigma(T) \setminus \cup_m \sigma^{(m)}$ is rotation invariant take any $\lambda \in \sigma(T) \setminus \cup_m \sigma^{(m)}$. If $\lambda \notin \cup_m \sigma_m$, then Proposition 12.15(2) implies that $\lambda\Gamma \subseteq \sigma(T)$ and we are done. So we will assume that $\lambda \in \sigma_m$ for some m. Let $Y_m := (h^{(m)} - g^{(m)})X$. Notice that Y_m is a band in X_m complemented to $X^{(m)}$, and therefore $X^{(m)}$ is isometrically isomorphic to X_m/Y_m. Since $TX_m \subseteq X_m$ and (as can be easily seen) $TY_m \subseteq Y_m$, the operator $T^{(m)}$ is similar to the quotient operator $\dot{T}_m$. If $(\lambda\Gamma) \cap \sigma_a(T|Y_m) \neq \emptyset$, then by Proposition 12.15(3) we have that $\lambda\Gamma \subset \sigma_a(T|Y_m) \subset \sigma_a(T)$. Therefore, it remains to consider the following two cases:

(a) $\lambda\Gamma \cap \sigma(T|Y_m) = \emptyset$. Then operators $(T - \lambda I)|Y_m$ and $\dot{T}$ - λI are invertible, and thus, by 12.18.0(1), operator $(T - \lambda I)|X_m$ is also invertible. This contradicts our assumption that $\lambda \in \sigma_m$. Consider the last case.

(b) $\lambda\Gamma \subset \sigma_r(T|Y_m)$. Assume that there exists a $\beta \in \lambda\Gamma \setminus \sigma_m$. Then operator $(T - \beta I)|Y_m$ is one-to-one but is not onto (since $\lambda\Gamma$ belongs to its spectrum) and the operator $(T - \beta I)|X_m$ is invertible. By 12.18.0(2), the operator $\dot{T}_m - \beta\dot{I}$ is onto but not one-to-one. However, this is impossible since $(\dot{T}_m)^m \in Z(X/X_m)$ and this case was refuted when we proved the inclusion $\sigma^{(m)} \subset \sigma(T)$. ■

12.19. Remark. Under the conditions of Proposition 12.18 the spectrum $\sigma^{(m)} = \sigma(T^{(m)})$ can be completely described. Indeed, as we showed the m-th power of the operator $T^{(m)}$ is the multiplication operator by a function $u_m \in Z(X)$, and therefore the spectrum of this operator $(T^{(m)})^m$ coincides with the range of the function u_m. Consequently by the spectral mapping theorem and by Proposition 12.15(1) we have that $\sigma^{(m)} = \{\lambda \in \mathbf{C} : \lambda^m = u_m(t) \text{ for some } t \in G^{(m)}\}$. This observation will be used repeatedly in Section 13.

12.19.1. Proposition. (E. Scheffold [Sch, Theorem V.4.4]) *Let T be a lattice* *homomorphism on a Banach lattice X. Then the spectrum $\sigma(T)$ is cyclic, that is, if $\lambda = re^{i\theta} \in \sigma(T)$, then $re^{in\theta} \in \sigma(T)$ for each $n \in \mathbf{N}$.*

Proof. Assume first that X is Dedekind complete and satisfies (C). Since T is positive, the functions u_m defined in Remark 12.19 are positive, and our claim follows from Proposition 12.18. The general case reduces to the previous one by considering the second conjugate operator T^{**} on the Dedekind complete Banach lattice X^{**}, and using the fact that $\sigma(T) = \sigma(T^{**})$. ■

Now we are ready to present the proof of Proposition 12.14.

Proof of Proposition 12.14. First we will prove that $\sigma(T)$ is rotation invariant. As we have already done several times, we can assume without loss of generality, that $\lambda \in \sigma_a(T)$ and also that $\lambda = 1$. Let $x_n \in X, \|x_n\| = 1$, satisfy $Tx_n - x_n \to 0$. Let us fix an $m \in \mathbf{N}$. By Frolik's invariant base theorem (see Corollary A.4) any fixed point of the mapping φ_T has a base of φ_T-invariant neighborhoods. From this and from the condition that $F_T^{(m)} \cap \operatorname{int} V_T^{(m)} = \emptyset$ we can easily infer that there exists a clopen neighborhood U_m of $F_T^{(m)}$ such that $\varphi_T(U_m) \subseteq U_m$ and $U_m \cap \operatorname{int} V_T^{(m)} = \emptyset$. Let η be a continuous mapping of $Q = Q_{X^{**}}$ onto K corresponding to the natural embedding of $C(K)$ into $Z(X^{**})$ (see 12.8). As $\eta \circ \varphi_{T^{**}} = \varphi_T \circ \eta$, we have the inclusion $F_{T^{**}}^{(m)} \subseteq Q_m := \eta^{-1}(U_m)$. We will proceed by considering two different cases. To begin with we assume that

$$\liminf_n \|\chi_{F_{T^{**}}^{(m)}} j x_n\| = 0.$$

Replacing $\{x_n\}$ by an appropriate subsequence we can assume additionally that $\lim_n \|\chi_{F_{T^{**}}^{(m)}} j x_n\| = 0$. Now for each $\alpha \in \Gamma$, we can (imitating the proof of Proposition 12.15) pick a unimodular function $g \in Z(X^{**})$ such that

$$\inf_n \|T^{**} g j x_n - \alpha g j x_n\| \le \frac{(2\pi + 1)}{m + 1}.$$

Consider the remaining case

$$\liminf_n \|\chi_{F_{T^{**}}^{(m)}} j x_n\| \ge c > 0. \tag{12.3}$$

Then

$$\liminf_n \|\chi_{U_m} x_n\| \ge c > 0.$$

Fix some $\alpha \in \Gamma$. The set $V_T^{(m)} \cap U_m$ is nowhere dense in U_m and (repeating the arguments from the proof of Theorem 12.11) we can find $g \in Z(X^{**})$ such that $\|\chi_{Q_m} g j x_n\| = \|\chi_{Q_m} j x_n\|$ and

$$|g(\varphi_{T^{**}}(t)) - \alpha g(t)| \le \frac{2\pi}{m+1} \quad \text{for } t \in Q_m.$$

Obviously, $\varphi_{T^{**}}(Q_m) \subseteq Q_m$ and thus $\varphi_{T^{**}}^{-1}(Q_m) \supset Q_m$. Let $g(t) := \alpha^{-n} g(\varphi_{T^{**}}^n(t))$ for $t \in \varphi_{T^{**}}^{-n}(Q_m) \backslash \varphi_{T^{**}}^{-(n-1)}(Q_m)$ $(n = 1, 2, \ldots)$. We can extend g (preserving continuity) to $H := \mathrm{cl}(\cup_{n=1}^{\infty} \varphi_{T^{**}}^{-n}(Q_m))$. Then $T^{**}\chi_H = \chi_H T^{**}$, and therefore $\|\chi_H j x_n\| \ge c$ and

$$\liminf_n \|T^{**}\chi_H j x_n - \alpha \chi_H j x_n\| \le \frac{2\pi}{m+1}.$$

Since m can be arbitrarily large and the constant c in (12.3) is independent of m (because $F_{T^{**}}^{(m)} \subset F_{T^{**}}^{(r)}$ for $m < r$) we have proved that $\Gamma \subseteq \sigma_a(T^{**}) (= \sigma_a(T))$.

Now take an arbitrary $T_1 \in Z(X)^{com}$ and consider $S := T_1 T$. For the operator S and each $f \in C(K)$ we have: $Sf = T_1 T f = T_1 (f \circ \varphi_T) T = (f \circ \varphi_T) T_1 T = (f \circ \varphi_T) S$. Therefore $F_S^{(m)} \cap \mathrm{int} V_S^{(m)} = \emptyset$, and hence $\sigma(S)$ is rotation invariant by the previous part of the proof.

To prove the converse part of Proposition 12.14 we again consider two cases. We begin assuming that there are a point $t \in F_T$ and a clopen neighborhood V of t such that

$$\varphi_T(V) = V \cap F_T. \tag{12.4}$$

Fix a positive $\varepsilon < 1/\|T\|$. Then for each $x \in X$ there exists an element $s(x) = \sum_{n=0}^{\infty} \varepsilon^n |T^n x|$. It is obvious that the closed ideal $X(s(x))$ generated by $s(x)$ is invariant under T and so we can consider the operator $T_x = T|X(s(x))$, the restriction of T to $X(s(x))$.

It follows from Theorem 3.17(1), Remark 3.19(1) and equality (12.4) that $\chi_V T_x^2 = \chi_V h_x T_x$ for some $h_x \in Z(X)$ satisfying $h_x(\varphi_T(q)) = h_x(q)$ for each $q \in V$. Moreover if $X(s(x)) \subseteq X(s(y))$ for some $y \in X$, then using the multiplicative representation we can see that $h_y = h_x$ on $supp(h_x)$. This forces the existence of $h \in Z(X)$ such that

$$\chi_V T^2 = \chi_V h T, \tag{12.5}$$

and

$$h(\varphi_T(q)) = h(q) \ \text{ for each } \ q \in V. \tag{12.6}$$

Observe that in view of the inclusion $V \cap F_T \subseteq E_{T^2}$ this function h cannot be identically zero on V, and hence we can pick a point $t_1 \in V$ at which $h(t_1) \neq 0$. Moreover, in view of (12.6) we can actually assume that $t_1 \in V \cap F_T$.

Take a characteristic function $g \in Z(X) = C(Q(X))$ such that $g(t_1) = 1$ and $supp(g) \subseteq V$. It is obvious that by diminishing (if necessary) the support of the function g we can additionally assume that the following conditions are satisfied:

- $|(gh)(q)| > \delta$ for some $\delta > 0$ and all $q \in supp(g)$.
- the set $\{(gh)(q) : q \in V \cap F_T\}$ is so small that it does not contain any circle $\gamma\Gamma$ with $\gamma > 0$.

Notice also that since $t_1 \in E_T \cap supp(g)$ we can find some $x \in X$ for which

- $(\chi_V gT)(x) \neq 0$.

We proceed by showing that the first and the third conditions imply that $r(gT) > 0$. To this end notice that for each $n \in \mathbf{N}$ we have

$$\chi_V(gT)^n x = u^{n-1}\chi_V gT(x), \tag{12.7}$$

where the function $u \in C(V)$ is defined by the formula $u(q) = g(\varphi_T(q)) \cdot h(\varphi_T(q))$ for $q \in V$. Thus (12.7) implies that $\|\chi_V(gT)^n x\| \geq \delta^{n-1}\|\chi_V gTx\|$, whence obviously $r(gT) > 0$.

Fix any $0 \neq \lambda \in \sigma_a(gT)$, which certainly exists since $r(gT) > 0$. Consequently there exists a normed sequence $\{x_n\} \subset X$ such that $\|gTx_n - \lambda x_n\| \to 0$. This implies in particular that $\|(gT)^2 x_n - \lambda^2 x_n\| \to 0$.

In view of (12.5) we have the identity $(gT)^2(x_n) = (g \circ \varphi_T)hgTx_n$, whence clearly $(g \circ \varphi_T)hx_n - \lambda x_n \to 0$, implying that $\lambda \in \sigma_a((g \circ \varphi_T)h)$. The operator $(g \circ \varphi_T)h$ is central, and so its approximate spectrum coincides with its spectrum and coincides with the range of the function generating this operator. That is, we have proved the equality $\sigma_a((g \circ \varphi_T)h) = \{(gh)(q) : q \in V \cap F_T\}$. In other words, we have established the identity $\sigma_a(gT) = \{(gh)(q) : q \in V \cap F_T\}$, and hence, in view of our assumption about the latter set, the spectrum $\sigma_a(gT)$ and consequently the spectrum $\sigma(gT)$ are not rotation invariant.

Consider the general case. It will be reduced to the previous one. Recall that by our assumption $F_T^{(m)} \cap int(V_T^{(m)}) \neq \emptyset$. Fix any $s \in F_T^{(m)} \cap int(V_T^{(m)})$, and let $p \in \mathbf{N}$, $p \leq m$ be the φ_T-period of the point s. By Frolik's invariant base theorem (see Corollary A.4) there is a clopen set V such that $s \in V$ and $\varphi^p(V) \subseteq V \subseteq int(V_T^{(m)})$. Then obviously $\varphi^p(V) = V \cap F_{T^p}$. This shows that operator T^p satisfies condition (12.4), and so by the preceding part there exists an $h \in Z(X)$ such that $\chi_V T^{2p} = h\chi_V T^p$, and also there exists a point $s_1 \in V \cap F_{T^p}$ such that $h(s_1) \neq 0$.

Recall that $Z(X) = C(Q(X))$, and so we can certainly find a function g such that $g(s_1) = g(\varphi_T(s_1)) = \cdots = g(\varphi_T^{p-1}(s_1)) = 1$.

Finally consider the following function $g_p := g \cdot (g \circ \varphi_T) \cdots (g \circ \varphi_T^{p-1})$. Again we can assume additionally that the set $\{(g_p h)(q) : q \in V \cap F_{T^p}\}$ contains no circle $\lambda\Gamma$ with $\lambda > 0$. Therefore, by the previous part of the proof we know that the spectrum $\sigma(g_p T^p)$ is not rotation invariant. A simple verification shows that $(gT)^p = g_p T^p$, so we can conclude that $\sigma(gT)$ is not rotation invariant either. ■

We conclude this section with a result about the spectrum of a d-homomorphism whose conjugate is also a d-homomorphism. This result is a partial extension of Proposition 12.18 to the case of $C(K)$-modules.

12.20. Proposition. *Let X be an order complete $C(K)$-module and $T \in dh(X)$ be such that $T^* \in dh(X^*)$. Let also either of the following two conditions hold.*

(1) *T is order continuous.*

(2) *X satisfies* (C).

Then $\sigma_m = \sigma^{(m)}$ for all $m \in \mathbf{N}$ and $\sigma(T) \setminus \bigcup_m \sigma^{(m)}$ is rotation invariant.

Proof. Assume first (1). Fix an arbitrary $x \in X$. Then $T|X(x)$ is an order continuous d-homomorphism of $X(x)$ into $X(Tx)$. Let us verify that the conjugate of this operator $T|X(x)$ is a d-homomorphism of $(X(Tx))^*$ into $(X(x))^*$. By Lemma 7.11(4), it will suffice to show that if $(x^* \square Tx)\,\mathrm{d}\,(y^* \square Tx)$, then $(T^*x^* \square x)\,\mathrm{d}\,(T^*y^* \square y)$. Denote by L the band in L_X^1 generated by all the measures $z^* \square Tx$, $z^* \in X^*$ and let

$A(z^* \,\square\, Tx) := T^*z^* \,\square\, x$. Then A is a linear continuous operator from L into L^1_X. If $h \in L^\infty_X$, then $Ah(z^* \,\square\, Tx) = T^*hz^* \,\square\, x = \gamma_{T^*}(h)(T^*z^* \,\square\, x) = \gamma_{T^*}(h)A(z^* \,\square\, Tx)$. Obviously $A \in dh(L, L^1_X)$ because $T^* \in dh(X^*)$. Now notice that in view of [Har] the mapping φ_T is injective on $supp(Tx)$, i.e., φ_T is a homeomorphism from E_T to K. By Frolik's Theorem (cited in 11.30) $F^{(m)}_T$ is clopen for each $m \in \mathbf{N}$. This clearly implies that $T^{(m)} = T_m$, $T^{(m)}$ commutes with T, and $\sigma^{(m)} \subset \sigma(T)$. The remaining part of the proof that the set $\sigma(T) \setminus \cup_m \sigma^{(m)}$ is rotation invariant may be obtained similar to the proof of Proposition 12.15 if we take into consideration that the set $E_T \setminus \mathrm{cl}(\cup_{m=1}^\infty F_T^{(m)})$ does not contain φ_T-periodic points (see also Remark 12.16).

Assume now (2). Since the operator T^{**} is order continuous, the previous part of our proof implies that mapping $\varphi_{T^{**}}$ is a homeomorphism from $E_{T^{**}}$ to Q. Therefore $\varphi^{-1}_{T^{**}}(F^{(m)}_{T^{**}}) = F^{(m)}_{T^{**}}$. Hence Remark 11.37 implies that

$$\eta^{-1}(intF_T^{(m)}) \subset F^{(m)}_{T^{**}} \quad \text{and} \quad \eta^{-1}(\varphi_T^{-1}(intF_T^{(m)})) \subset \varphi^{-1}_{T^{**}}(F^{(m)}_{T^{**}}) = F^{(m)}_{T^{**}}.$$

Taking into consideration that $\varphi_T \circ \eta = \eta \circ \varphi_{T^{**}}$, we obtain $\varphi_T^{-1}(intF_T^{(m)}) = intF_T^{(m)}$ for all $m \in \mathbf{N}$, whence $\sigma_m = \sigma^{(m)}$ and $T^{(m)}T = TT^{(m)}$. Now we can proceed exactly as in the proof of Proposition 12.15(2). ∎

Recall that an operator $T \in dh(X)$ on a Banach lattice is said to be quasi-invertible [Har], provided $T^* \in dh(X^*)$, T is one-to-one, and $\{TX\}^{dd} = X$.

Since each quasi-invertible operator is order continuous [Har], the next corollary follows immediately from the previous proposition.

12.20.1. Corollary. (Arendt and Hart [AH]) *Let X be an arbitrary Dedekind complete Banach lattice and let $T : X \to X$ be a quasi-invertible operator on X. Then $\sigma(T) \setminus \cup_m \sigma^{(m)}$ is rotation invariant.*

The further improvements of the Arendt-Hart theorem will be given in Theorem 13.14 and Proposition 13.15.

13. Spectrum of d-endomorphisms on Banach lattices

Recall one definition from the theory of Banach lattices. Let X be a Banach lattice. For each $x \in X$ consider the expression

$$\ell_X(x) = \inf_{\{x_\alpha\}} \sup_\alpha \|x_\alpha\|,$$

where the infimum is taken over all nets $\{x_\alpha\} \subset X_+$ such that $x_\alpha \uparrow |x|$. This functional $\ell_X(x)$ is a Riesz semi-norm on X and it is called the Lorentz seminorm [LZ1]. Obviously $\ell_X(x) \leq \|x\|$ and $\ell_X(x) = \|x\|$ if and only if X satisfies (C).

In this section we will show that in the case of Banach lattices, the conclusion of Theorem 12.11 that the spectrum $\sigma(T)$ is rotation invariant remains true if we replace the assumption that X satisfies (C) by a considerably weaker assumption that

$$\ell_X(\cdot) \text{ is a norm.} \tag{$\star$}$$

For any Banach lattice satisfying ($\star$), we will describe the spectrum of any order continuous d-endomorphism on this lattice. It is appropriate to note at this point that the class of Banach lattices X for which ℓ_X is a norm is very large; for example, any Banach function space belongs to this class. We will also show that any d-homomorphism T with independent powers on an arbitrary Banach lattice has a rotation invariant spectrum provided $T^* \in dh(X^*)$.

13.1. Theorem. *Let X be an arbitrary Dedekind complete Banach lattice satisfying ($\star$). Let $T \in dh(X)$ be order continuous. Assume finally that $\sigma(T) = \sigma_i \cup \sigma_e$, where the sets σ_i, σ_e are closed, $\sigma_i \subset int(D_r)$ and $\sigma_e \cap D_r = \emptyset$ for some $r \geq 0$. Then there exists a band X_i in X such that $TX_i \subset X_i$, $\sigma(T|X_i) = \sigma_i$ and $\sigma(P_Y T|Y) = \sigma_e$, where $Y = (X_i)^d$ is the disjoint complement to X_i and P_Y is the band projection of X onto Y.*

Proof. Assume first that X satisfies (C), and denote by X_i the spectral subspace of the operator T corresponding to σ_i. We claim that X_i is an ideal in X. Indeed, if $x \in X_i$ and $|y| \leq |x|$, then $\limsup_n \|T^n y\|^{\frac{1}{n}} \leq \limsup_n \|T^n x\|^{\frac{1}{n}} \leq r$, whence $y \in X_i$

and hence X_i is an ideal in X. Now we will show that X_i is actually a band. Let $x_\alpha \in X_i$ and $x_\alpha \uparrow x \in X$. Choose a positive δ small enough to guarantee that $\sigma_e \cap D_{r+\delta} = \emptyset$. Then $\|T^n x_\alpha\| \le c(r+\delta)^n \|x_\alpha\|$, where $c > 0$ is independent of n and α. Since T is order continuous and X satisfies (C), we have $\|T^n x\| \le c(r+\delta)^n \|x\|$, and thus $x \in X_i$, that is, X_i is, indeed, a band in X. Let X_e denote the spectral subspace corresponding to σ_e. The mapping $x \to P_Y x$, $x \in X_e$, is a one–to–one linear operator from X_e onto Y. This isomorphism establishes a similarity between operators $T|X_e$ and $P_Y T|Y$. Therefore $P_Y T|Y$ is a d-automorphism of the lattice Y, and hence $\sigma(P_Y T|Y) = \sigma_e$.

Return to the general case. Recall that condition (C) is always fulfilled in the conjugate space, and so we can apply the case considered above to the operator T^{**}. The corresponding bands in X^{**} will be denoted by $X_i^{(**)}$ and $Y^{(**)}$, respectively. Let $Q_2 = supp(Y^{(**)})$ and let $K_2 = \eta(Q_2)$, where the mapping η between the Stonean spaces of X^{**} and X was introduced in 12.8. In the present proof we will denote by Q (resp. K) the Stonean space of the former (resp. latter) Banach lattice.

We claim that $K_2 = int(K_2)$. If not, then there exists $x \in X$ such that $supp(x) \cap K_2$ is a nonempty nowhere dense subset of K. Let P_e denote the spectral projection onto the spectral subspace X_e of X and let $y = P_e x$. By the first part of our proof $supp(jy) \cap Q_2 = supp(jx) \cap Q_2$, and therefore, in view of the identity $supp(jx) = \eta^{-1}(supp(x))$, we have that $supp(y) \cap K_2 = supp(x) \cap K_2$.

Since $0 \notin \sigma_e$ the operator $T|X_e$ is invertible, and therefore for each $k \in \mathbf{N}$ we can find $y_k \in X_e$ such that $T^k y_k = y$. Multiplying T by an appropriate constant, we can assume for convenience that $r = 1$, and hence $r((T|X_e)^{-1}) < 1$. Consequently there exists a $\delta > 0$ such that $\|y_k\| \le (1-\delta)^k \|y\|$ for all sufficiently large k. The mapping φ_T is open and $\varphi_T(K_2) = K_2$ (since $\varphi_{T^{**}}(Q_2) = Q_2$). Consequently, the set $supp(y_k) \cap K_2 = \varphi_T^{-k}(supp(y)) \cap K_2$ is a nowhere dense subset of K. Thus for each $k \in \mathbf{N}$, there exists a net $\{y_{k,\alpha}\} \subset X$ such that $supp(y_{k,\alpha}) \cap K_2 = \emptyset$ and $y_{k,\alpha} \uparrow_\alpha |y_k|$. Order continuity of T implies that $|T^k y_{k,\alpha}| \uparrow_\alpha T^k|y_k| = |T^k y_k| = |y|$. It is clear that $j y_{k,\alpha} \in X_i^{(**)}$ and since $r(T^{**}|X_i^{(**)}) < 1$, we get the following inequalities:

$$\|T^k y_{k,\alpha}\| \le \|y_{k,\alpha}\| \le \|y_k\| \le (1-\delta)^k \|y\|$$

for all sufficiently large k. Hence $\ell_X(y) \le (1-\delta)^n$ for all $n \in \mathbf{N}$, i.e., $\ell_X(y) = 0$; whence

$y = 0$, a contradiction. This proves that $K_2 = \text{ int}(K_2)$.

Let $Y := \{x \in X : supp(x) \subset K_2\}$ and $X_i = (Y)^d$. We claim that X_i coincides with the spectral subspace of T corresponding to σ_i. Indeed, if an element $x \in X_i$, then $supp(jx) \cap Q_2 = \emptyset$ and thus, $jx \in X_i^{(**)}$. Conversely, if $x \notin X_i$, then $supp(jx) \cap Q_2 \neq \emptyset$ and $jx \notin X_i^{(**)}$, i.e., in this case x has a nonzero projection on X_e. This completes the proof of this theorem. It is interesting to note that in the general case, i.e., in the absence of (C), we first construct the band Y and afterwards $X_i = Y^d$, as opposed to the case with (C). ■

We need a remark of a technical nature.

13.2. Remark. Under the assumptions and notations of the previous theorem $supp(X_e) \subset V := cl(\bigcup_{m=1}^{\infty} \varphi_T^{-m}(K_2))$. Indeed, the projection $P_V = \chi_V$ commutes with T, and therefore it commutes with the spectral projection on X_e. Thus $P_V y \in X_e$ for each element $y \in X_e$. But $supp(y - P_V y) \cap K_2 = \emptyset$, and hence $y - P_V y \in X_i$. This gives $y = P_V y$.

In the next lemma we retain the notations used in the previous theorem, though now X is an arbitrary Banach lattice and $T \in dh(X)$ is arbitrary, too.

13.2.1. Lemma. *For each integer m the operator $(T|X_e)^m P_e \in dh(X)$. In particular, taking $m = 0$ we have that the spectral projection P_e preserves disjointness.*

Proof. Without loss of generality we can assume that T is order continuous and X satisfies (C) (otherwise we will replace X by X^{**} and replace the operators by their second conjugate). Then by Theorem 13.1, $P_Y T|Y$ is a d-automorphism of the band Y. We denote by S_0 the inverse $(P_Y T|Y)^{-1}$ and by P_n, $n \in \mathbf{N}$, the band projection generated by the clopen set $\varphi_T^{-n}(K_2) \setminus \varphi_T^{-n+1}(K_2)$. It is plain to see that $P_n(T|X_e)^m P_e = P_n T^{n+m} S_0^n P_Y$, and $P_Y P_e = P_Y$. This and Remark 13.2 imply that $(T|X_e)^m P_e \in dh(X)$. ■

13.3. Proposition. *Let X be a Banach lattice satisfying $(\star)$ and $T \in dh(X)$. Let $\sigma(T) = \sigma_i \cup \sigma_e$, where $\sigma_i \subset int(D_r)$ and $\sigma_e \cap D_r = \emptyset$ for some $r > 0$. Finally, let X_e be the spectral subspace of T corresponding to σ_e. Then there exist $y \in X_e$ and $n \in \mathbf{N} \cup \{0\}$*

such that the set $A = A(n,y) = \eta_{T^n y}(\varphi_{T^{**}}^{-n}(supp(jy) \cap Q_2))$ *has a nonempty interior in* $K_{T^n y}$. *(Recall that the mapping* $\eta_{T^n y}$ *and the set* $K_{T^n y}$ *were introduced in* 12.8.)

Proof. Let us assume, contrary to what we claim, that

$$int A(n,y) = \emptyset \quad \forall\, y \,\&\, n. \tag{13.1}$$

Now fix any $0 \neq y \in X_e$. For each $k \in \mathbf{N}$ there exists $y_k \in X_e$ such that $T^k y_k = y$. Again, as in the proof of Theorem 13.1, we can assume for convenience that $r = 1$. Then for all k large enough we have $\|y_k\| \leq \|y\|$. We claim the existence of a net $y_{\alpha,k}$ in X satisfying the following three properties.

(1) $supp(j y_{\alpha,k}) \cap Q_2 = \emptyset$.

(2) $y_{\alpha,k} \uparrow_{\alpha} |y_k|$.

(3) $T^k y_{\alpha,k} \uparrow_{\alpha} |T^k y_k| = |y|$.

It is worthwhile to notice that unlike the proof of Theorem 13.1 condition (3) is not a consequence of (2) since T is not assumed to be order continuous. Nevertheless, the additional property (3) can still be obtained if one uses (13.1). Here are the details.

The set $A(0, y_k) = \eta_{y_k}(supp(j y_k) \cap Q_2)$ is nowhere dense in K_{y_k}. Let $\mathcal{U}$ be the collection of all pairs $U = (V, W)$ of open neighborhoods of the set $A(0, y_k)$ such that $cl(V) \subset W$. We say that $U_1 = (V_1, W_1) \prec U_2 = (V_2, W_2)$ if $cl(W_2) \subset V_1$. For each $U = (V, W) \in \mathcal{U}$ we can find an element y_U such that†

(a) $y_U \leq |y|$,

(b) $y_U = |y|$ outside W, and

(c) $y_U = 0$ on V.

Obviously the net $\{y_U\}$ satisfies (1) and (2) above. Operator $T^k|X(y_k)$ is a d-homomorphism of $X(y_k)$ onto $X(T^k y_k) = X(Ty)$. Let ψ be the corresponding mapping of K_y onto K_{y_k}. Then we have $\psi \circ \eta_y = \eta_{y_k} \circ \varphi_{T^{**}}$, and consequently $\psi^{-1}(A(0, y_k)) = A(k, y_k)$. The last set is nowhere dense in K_y by (13.1). This clearly implies that the net $\{T^k y_U\}$ satisfies (3).

---

† Recall that by the Krein-Kakutani theorem the ideal X_x is isomorphic to $C(K_x)$ for each $x \in X$.

Using (1)–(3) and imitating the concluding part of the proof of Theorem 13.1 we can show that $\ell_X(y) = 0$. This implies that $y = 0$, a contradiction. ∎

Now we are ready to prove the main result of this section.

13.4. Theorem. *Let X be a Banach lattice satisfying $(\star)$ and $T \in dh(X)$. If $T^m \delta\ T^n$ for any distinct $m, n \in \mathbf{N} \cup \{0\}$, then $\sigma(T)$ is rotation invariant.*

Proof. Assume that $1 \in \sigma_a(T)$ but $\Gamma \not\subset \sigma(T)$. Then we can find a $\delta > 0$ and $m \in \mathbf{N}$ such that the set $\sigma(T) \cap \{\lambda \in \mathbf{C} : 1 - 2\delta \leq |\lambda| \leq 1 + 2\delta\}$ belongs to an angle with the vertex at the origin and of measure less than $2\pi(1 - \frac{1}{m+1})$. Certainly the same is true for T^{**}. As $T^{**} : X^{**} \to X^{**}$ is order continuous we have, by Proposition 12.18(1), that $\zeta_k \sigma^{(k)} \subseteq \sigma^{(k)}$, where $\zeta_k = \exp(2\pi i/k)$ and $\sigma^{(k)} := \sigma((T^{**})^{(k)})$. Now we clearly have that

$$\sigma(T^{**}) \cap \{\lambda \in \mathbf{C} : 1 - 2\delta \leq |\lambda| \leq 1 + 2\delta\} \subseteq \cup_{k=1}^{m} \sigma^{(k)}.$$

By Remark 12.19, applied to the operator T^{**}, we get

$$\sigma^{(k)} = \{\lambda : \lambda^l \in \{u_l(t) : t \in G_{T^{**}}^{(l)}\}\},$$

where $u_l (= u_l(T^{**}))$ is a corresponding function from $Z(X^{**})$.

We will need the following subset of $G_{T^{**}}^{(l)}$.

$$G_\delta^{(l)} = \mathrm{cl}\{t \in G_{T^{**}}^{(l)} : (1 - \delta)^l < |u_l(t)| < (1 + \delta)^l\}.$$

Also let $F_\delta^{(m)} = \cup_{l=1}^{m} G_\delta^{(l)}$. And consider now the following subset of X:

$$X_\delta := \{x \in X : \ supp(jx) \cap F_{T^{**}}^{(m)} \subset F_\delta^{(m)}\}.$$

It is easy to see that X_δ is a closed ideal in X. Let us verify that $TX_\delta \subset X_\delta$. Take any $x \in X_\delta$. If $t \in supp(T^{**}jx) \cap F_{T^{**}}^{(m)}$, then $\varphi_{T^{**}}(t) \in supp(jx) \cap F_{T^{**}}^{(m)} \subset F_\delta^{(m)}$. To simplify subsequent formulas let us introduce the operator $A := (T^{**})^{(m)}$. The chain of the equalities

$$u_l A = A^m A = A A^m = A u_l = (u_l \circ \varphi_{T^{**}}) A$$

shows that$u_l \circ \varphi_{T^{**}} \equiv u_l$ on $F_{T^{**}}^{(m)}$, i.e., $u_l(t) = u_l(\varphi_{T^{**}}(t))$. Consequently $t \in F_\delta^{(m)}$ and this proves that $Tx \in X_\delta$.

We claim that

$$\sigma(T|X_\delta) \cap (1-2\delta)\Gamma = \sigma(T|X_\delta) \cap (1+2\delta)\Gamma = \emptyset. \tag{13.2}$$

We start by showing that

$$\sigma_a(T|X_\delta) \cap (1-2\delta)\Gamma = \emptyset.$$

If not, then there are β, $|\beta| = 1-2\delta$, and $x_n \in X_\delta$ such that $\|x_n\| = 1$ and $Tx_n - \beta x_n \to 0$. Since the circle $(1-2\delta)\Gamma$ does not belong to $\sigma_a(T)$, the proof of Proposition 12.15(3) shows that there exists some l, $1 \le l \le m$, such that

$$T^{**} g^{(l)}_{T^{**}} j x_n - \beta g^{(l)}_{T^{**}} j x_n \to 0$$

and $y_n := g^{(l)}_{T^{**}} j x_n \not\to 0$. On the other hand, Remark 12.19 implies that the l-th power of the operator $(T^{**})^{(l)}$ coincides with multiplication at u_l, and therefore clearly $u_l y_n - \beta^l y_n \to 0$ as $n \to \infty$. However, $(1-\delta)^l < |u_l| < (1+\delta)^l$ on $supp(y_n)$ and $|\beta|^l = (1-2\delta)^l$. Therefore, $\|u_l y_n - \beta^l y_n\| \ge [(1-\delta)^l - (1-2\delta)^l] \cdot \|y_n\| \not\to 0$, a contradiction.

Now, if we show that there exists at least one point $\xi \in (1-2\delta)\Gamma$ for which the equality $(T - \xi I)X_\delta = X_\delta$ holds, then (using that the topological boundary of the spectrum $\sigma(T)$ belongs to $\sigma_a(T)$) it will imply that $\sigma(T|X_\delta) \cap (1-2\delta)\Gamma = \emptyset$. Let $\xi \in (1-2\delta)\Gamma \backslash \sigma(T)$ and $x \in X_\delta$. Then there is a point $y \in X$ such that $Ty - \xi y = x$. Let $P_\delta \in Z(X^{**})$ be a projection on X^{**} corresponding to the set $F^{(m)}_{T^{**}} \backslash F^{(m)}_\delta$. Then $P_\delta T^{**} y - \xi P_\delta y = P_\delta x = 0$. As the set $F^{(m)}_{T^{**}} \backslash F^{(m)}_\delta$ is $\varphi_{T^{**}}$-invariant, we have

$$P_\delta T^{**} P_\delta y = P_\delta (P_\delta \circ \varphi_{T^{**}}) T^{**} y = P_\delta T^{**} y,$$

whence $P_\delta T^{**} P_\delta y - \xi P_\delta y = 0$. This implies that $P_\delta y = 0$, since otherwise the point $\xi \in \cup_{l=1}^m \sigma^{(l)}_{T^{**}} \subset \sigma(T^{**}) = \sigma(T)$ contrary to our assumption. Thus, indeed, $P_\delta y = 0$, and we can conclude that $y \in X_\delta$.

The equality $\sigma(T|X_\delta) \cap (1+2\delta)\Gamma = \emptyset$ can be verified similarly.

Now we claim that

$$\sigma(T|X_\delta) \cap \{\lambda \in \mathbf{C} : 1-\delta \le |\lambda| \le 1+\delta\} \ne \emptyset. \tag{13.3}$$

Fix an $x \in X$ such that $supp(jx) \cap F_\delta^{(m)} \neq \emptyset$.

Replacing (if necessary) x by $\sum_{n=0}^{\infty} c^n |T^n x|$, where $0 < c < \|T\|^{-1}$, we may additionally assume that $Tx \leq \gamma x$ for some positive constant γ, i.e., $TX_x \subset X_x$ and that $T^{**}(X^{**})_{jx} \subset (X^{**})_{jx}$. We will assume further that the ideals X_x and $(X^{**})_{jx}$ are represented as spaces $C(K_x)$ and $C(supp(jx))$, respectively, the functions x and jx being represented as the constant one functions. Take $p = m!$. By Remark 12.19, there exists a point $q \in F_\delta^{(m)}$ such that $(1-\delta)^p < |[(T^{**})^p jx](q)| < (1+\delta)^p$. Let $s = \eta_x(q) \in K_x$. Then $\lambda := (T^p x)(s) = [(T^{**})^p jx](q)$. Pick an open neighborhood V of s in K_x such that $(1-\delta)^p < \lambda < (1+\delta)^p$ for all $t \in V$. Let $g \in C(K_x)$ be such that $supp(g) \subset V$ and $g(s) = 1$. Letting $y = gx$, we see that $y \in X_\delta$ and $q \in supp(jy) \cap F_\delta^m$.

To establish (13.3) it will suffice to show that $\lambda^p \in \sigma(T^p|X_\delta)$. Assume, contrary to what we claim, that $T^p - \lambda^p I$ is invertible on X_δ. Let us fix an $\varepsilon > 0$ and pick a neighborhood V_ε of s in K_x such that the oscillation of the function $T^p x$ on V_ε is less than ε. Consequently we can find a function $h \in C(K_x)$ such that $\|h - \mathbf{1}\| < 2\varepsilon$ and

$$(hT^p x)(s') = (hT^p x)(s) \quad \forall\, s' \in V_\varepsilon. \tag{13.4}$$

Finally, introduce the operator $\tilde{T} = hT^p$. Clearly $\|\tilde{T} - T^p\| < 2\varepsilon\|T^p\|$, and hence $\tilde{T} - \lambda^p I$ is invertible on X_δ for small enough ϵ. Consequently there exists $z \in X_\delta$ for which $\tilde{T}z - \lambda^p z = y$. Now, in view of (13.4), $j(\tilde{T}z - \lambda^p z) \equiv 0$ on the set $\eta_x^{-1}(V_\varepsilon)$ which contains q. But on the other hand, $q \in supp(jy)$, a contradiction.

From (13.2) and (13.3) it follows that in X_δ there exists a nonzero spectral subspace V such that $\sigma(T|V) \subset \{\lambda : 1 - 2\delta < |\lambda| < 1 + 2\delta\}$. Let $y \in V$, $y \neq 0$. Proposition 13.3 implies that for some $k \in \mathbf{N}$

$$\operatorname{int} \eta_y(\varphi_{T^{**}}^{-k}(F_{T^{**}}^{(m)} \cap supp(jy))) \neq \emptyset. \tag{13.5}$$

However, in view of Proposition 11.36, condition (13.5) contradicts the independence of powers of the operator T. This completes the proof of the theorem. ∎

13.4.1. Corollary. *Let X be a Banach function space and $T \in dh(X)$. If $|T|^n \wedge |T|^m = 0$ for any $n \neq m$, then $\sigma(T)$ is rotation invariant.*

We refer to Proposition 12.14 which explains that the condition of disjointness of powers of T is (in some sense) necessary.

Now we are going to show that condition $(\star)$ in Theorem 13.4 cannot be omitted.

13.5. Proposition. *There exist a separable Banach lattice X and an operator $T \in dh(X)$ such that $T^m \delta\ T^n$ for any distinct $m, n \in \mathbf{N} \cup \{0\}$ and whose spectrum $\sigma(T) = \{0\} \cup \{1\}$, that is, $\sigma(T)$ is not rotation invariant.*

Proof. Let ψ_0 denote a function on [0, 1] defined as follows: $\psi_0(t) = 2t$ for $t \in [0, \frac{1}{2}]$ and $\psi_0(t) = 2 - 2t$ for $t \in [\frac{1}{2}, 1]$. Using this function ψ_0, we define a mapping $\psi : [0,1] \longrightarrow [0,1]$ by letting $\psi(0) = 0$ and letting $\psi(t) = \frac{1}{2^n}\psi_0(2^n t - 1)$ for $t \in [\frac{1}{2^n}, \frac{1}{2^{n-1}}]$, $n \in \mathbf{N}$.

It is easy to see that ψ has the following properties.

(1) For each $n \in \mathbf{N}$ the point $x = 0$ is a unique fixed point of the mapping ψ^n.

(2) For each $n \in \mathbf{N}$ the set $\psi^{-n}(0)$ is countable.

(3) The set $\cup_{n=1}^{\infty} \psi^{-n}(0)$ is dense in [0,1].

(4) For each $t \in \psi^{-n}(0)$, there is a sequence $t_p \in \psi^{-(n+1)}(0) \backslash \psi^{-n}(0)$ such that $t_p \to t$.

The property (1) is obvious while (2) and (3) follow from the observation that $\cup_{n=1}^{\infty} \psi^{-n}(0)$ is the set of all dyadic points in [0, 1]. Let us verify (4). Since $\psi^{-n}(0)$ is closed and countable, it is enough to assume that t is an isolated point of this set. If m is large enough, then we can find $t_p \in [0,1]$ such that $t_p \longrightarrow t$ and $\psi^n(t_p) = \frac{1}{2^{m+p}}$. It is obvious that $\{t_p\}$ is as desired.

Introduce the following functional on $C[0,1]$:

$$|||f||| = \sup_{n \in \mathbf{N}} \{ \frac{1}{n!} \max_{t \in \psi^{-n}(0)} |f(t)| \}. \tag{13.6}$$

It is plain to see that $|||\cdot|||$ is a lattice seminorm and (3) implies that $|||\cdot|||$ is, in fact, a norm.

We denote by X the completion of C[0, 1] with respect to the norm $|||\cdot|||$. Thus X is a separable Banach lattice.

Letting $Tf = f \circ \psi, f \in C[0,1]$, we have $|||Tf||| \leq |||f|||$, and therefore operator T admits an extension still denoted by T to a continuous d-homomorphism on X.

Similarly, the functional $\alpha : f \mapsto f(0), f \in C[0,1]$ admits an extension to a continuous functional (still denoted by) $\alpha \in X^*$. We denote by X_0 the zero-hyperplane of α. Thus, $X = X_0 \oplus X_1$ where X_1 is the one-dimensional subspace generated by the constant one function. Obviously $TX_1 = X_1$, $TX_0 \subset X_0$ and $\sigma(T|X_1) = \{1\}$. By the very definitions of $|||\cdot|||$ and T, we have

$$|||T^n f||| \leq \frac{1}{n!}|||f|||$$

if $f \in C[0,1]$ and $f(0) = 0$. Therefore, $r(T|X_0) = 0$, and consequently $\sigma(T) = \{0\} \cup \{1\}$.

Let us verify that $T^m \delta\, T^n$ for any $m \neq n \in \mathbf{N} \cup \{0\}$. Algebra $C[0,1]$ is isometrically embedded into center $Z(X)$ of the lattice X, and to this embedding there corresponds a mapping η of the maximal ideal Gelfand space K of the algebra $Z(X)$ onto [0, 1]. The lattice X has a quasi-interior point $\mathbf{1}$ and $T\mathbf{1} = \mathbf{1}$. Therefore to the operator T there corresponds a continuous mapping φ of K into itself.

Assume, contrary to what we claim, that $T^p\, \bar{\delta}\, T^q$ for some $p \neq q \in \mathbf{N} \cup \{0\}$. Proposition 11.36 (or a criterion for the dependence of powers of d-homomorphisms in a Banach lattice with a quasi-interior point, which will be mentioned in 14.12), property (1) of the mapping ψ and equality $\psi \circ \eta = \eta \circ \varphi$ show that there exist $x \in X$ and $n \in \mathbf{N}$ such that $|||x||| = 1$ and $supp_{[0,1]} x \subset \psi^{-n}(0)$. Let $0 < \zeta < \frac{1-2\zeta}{(n+1)!}$. Pick up a $g \in C[0,1]$ such that $|||x - g||| < \zeta$. Since $supp(x) \subset \psi^{-n}(0)$ and in view of (13.6), we can find a point $t \in \psi^{-n}(0)$ such that $|g(t)| \geq 1 - \zeta$. Therefore, by property (4) of the mapping ψ, we can find $t_1 \in \psi^{-(n+1)}(0) \backslash \psi^{-n}(0)$ such that $|g(t_1)| \geq 1 - 2\zeta$. Let $h \in C[0,1], \|h\|_\infty = h(t_1) = 1$ and $supp(h) \cap \psi^{-n}(0) = \emptyset$. Then $|||hg||| \geq \frac{1-2\zeta}{(n+1)!}$. On the other hand, since $supp(x) \subset \psi^{-n}(0)$, we have $|||hg||| \equiv |||h(g-x)||| \leq \zeta < \frac{1-2\zeta}{(n+1)!}$, a contradiction. ■

Theorem 13.1 allows us to improve Proposition 12.18 by replacing condition (C) by a weaker condition ($\star$). We omit the proof since it repeats the arguments of Proposition 12.18, substituting Theorem 13.1 for Proposition 12.15.

13.6. Proposition. *Let X be a Dedekind complete Banach lattice satisfying $(\star)$ and let $T \in dh(X)$ be order continuous. Then for each $m \in \mathbf{N}$ the inclusion $\sigma^{(m)} \subseteq \sigma(T)$ holds, and $\sigma(T) \backslash \cup_{m=1}^{\infty} \sigma^{(m)}$ is a rotation invariant subset of* $\mathbf{C}$.

13.6.1. Assume that all the conditions of the previous proposition are fulfilled. Consider a maximal clopen subset Q_h of $Q = Q(X)$ for which the restriction $\varphi_T | Q_h$ is a homeomorphism of Q_h onto itself. (Q_h, which may be empty, exists since K is extremally disconnected.)

If we let $X_2 = \{x \in X : \; supp(x) \subset Q_h\}$, $X_1 = \{X_2\}^d$ and $P = \chi_{Q_h}$, then operator $PT|X_2$ is quasi-invertible on X_2, $TX_1 \subset X_1$ and $\sigma(T|X_1)$ is a disc in view of Theorem 13.1 and Proposition 13.6. Indeed, by Proposition 13.6, the set $\sigma(T|X_1)$ is rotation invariant, since otherwise the set $G_T^{(m)} \cap (Q \backslash Q_h) \neq \emptyset$ for some $m \in \mathbf{N}$. But this contradicts the definition of Q_h since φ_T is a homeomorphism of $G_T^{(m)} \cap (Q \backslash Q_h)$ onto itself. In a similar way, using Theorem 13.1, we can conclude that $\sigma(T|X_1)$ is connected. Consequently $\sigma(T|X_1)$ is a disc.

13.7. Proposition. *Under the conditions of Corollary* 13.6, *the identity* $\sigma(T) = \sigma(T|X_1) \cup \sigma(PT|X_2)$ *holds.*

Proof. The inclusion $\sigma(T) \subset \sigma(T|X_1) \cup \sigma(PT|X_2)$ is a standard fact mentioned in 12.18.0(1): if the restriction of an operator to a subspace and the corresponding quotient operator are invertible, then the operator itself is invertible on the whole space.

Let $\lambda \notin \sigma(T)$. Two cases are possible.

(1) $\lambda\Gamma \cap \sigma(T) = \emptyset$. Then by Theorem 13.1, $r(T|X_1) < |\lambda|$, i.e., $\lambda \notin \sigma(T|X_1)$ and by the three spaces theorem (i.e., again by 12.18.0(1)) $\lambda \notin \sigma(PT|X_2)$.

(2) $\lambda\Gamma \cap \sigma(T) \neq \emptyset$. Then by Proposition 13.6, there exists an $m \in \mathbf{N}$ such that $\lambda\Gamma \cap \sigma(I - \sum_{l=1}^{m} h^{(l)}(T)) = \emptyset$. Therefore applying (1) to the operator $T - \sum_{l=1}^{m} h^{(l)}T$, we have that $\lambda \notin \sigma(P(T - \sum_{l=1}^{m} h^{(l)}T)|X_2)$. As soon as we verify that $\lambda \notin \sigma(Ph^{(l)}T|h^{(l)}X_2)$ for each $l = 1, 2, \dots, m$ we can conclude that $\lambda \notin \sigma(PT|X_2)$, and by 12.18.0(1), $\lambda \notin \sigma(T|X_1)$. The definitions of P and Q_h imply that $Ph^{(l)} = g^{(l)}$ and $h^{(l)}X_2 = g^{(l)}X$.

Therefore $Ph^{(l)}T|h^{(l)}X_2 = T^{(l)}$. It remains to notice that, by Proposition 13.6, $\sigma(T^{(l)}) = \sigma^{(l)} \subseteq \sigma(T)$, whence $\lambda \notin \sigma^{(l)}$. ■

The previous proposition gives a complete description of the spectrum of any order continuous d-homomorphism on Banach function spaces. However, the interesting and hard question of evaluation of the spectral radius is left open (cf. [Lat]).

13.8. Our next result shows that the developed technique is powerful enough to enable us to obtain rather useful information about the spectrum $\sigma(T)$ even without assuming that T is order continuous and that X satisfies (C) or ($\star$). (Cf. Propositions 12.15, 13.6 and 13.7.)

13.9. Theorem. *Let X be a Dedekind complete Banach lattice and $T \in dh(X)$. The following conditions* (1) *and* (2) *are equivalent.*

(1) *The exists a positive number r such that the circle $r\Gamma$ splits the spectrum $\sigma(T)$, that is, there are two non-empty subsets σ_i and σ_e of $\sigma(T)$ such that $\sigma(T) = \sigma_i \cup \sigma_e$, $\sigma_i \subset int(D_r)$ and $\sigma_e \cap D_r = \emptyset$.*

(2) *There exists a subset K_2 of $Q(X)$ satisfying the next five properties:*

- (a) *K_2 is a retract of $Q(X)$.*
- (b) *$r(T|X_0) < r$, where X_0 denotes the closed ideal in X generated by all $x \in X$ with $supp(x) \cap K_2 = \emptyset$.*
- (c) *The quotient operator $\dot{T}$ is a d-automomorphism of X/X_0 and $\sigma(\dot{T}) = \sigma_e$.*
- (d) *The quotient space X/X_0 is Dedekind complete and $Z(X/X_0)$ coincides with $C(K_2)$.*
- (e) *The restriction $\varphi_T|K_2$ is a homeomorphism of K_2 onto itself.*

Proof. (1) $\Rightarrow$ (2). Let $K_2 = \eta(Q_2)$, where Q_2, K_2 and η are the objects which were defined in the proof of Theorem 13.1. We want to verify the properties (a)–(e).

(a) By Lemma 13.2.1 the spectral projection P_e is a d-homomorphism. The corresponding mapping φ_e generated by P_e is a retraction of the set $supp(X_e)$ onto K_2, where $X_e = P_eX$ is the spectral subspace of X. Since $supp(X_e)$ is a clopen subset of $Q(X)$ we see that K_2 is also a retract of $Q(X)$. This implies in particular, that K_2 is extremally disconnected as a retract of extremally disconnected space.

Property (b) follows from the definition of K_2.

Property (c) follows from the fact that the canonical embedding $x \to \dot{x}$ of X onto X/X_0 establishes an isomorphism between $X_e = P_e X$ and X/X_0, and the fact that the operator $T|X_e$ is similar to $\dot{T}$. (See the arguments used in the proof of Theorem 13.1.)

(d) It is plain to see that X/X_0 is an operator $C(K_2)$-module. If $0 \neq \dot{x} \in X/X_0$ and x is its (unique) preimage in X_e, then as was shown in the proof of Theorem 13.1 $supp_{Q_2}(j\dot{x}) = supp_{Q(X^{**})}(jx) \cap Q_2$. This implies that $supp_{K_2}(\dot{x}) = supp_{Q(X)}(x) \cap K_2$. Therefore by Proposition 11.20 and Remark 11.19.1, the space X/X_0 is a Dedekind complete Banach lattice and its center coincides with $C(K_2)$.

(e) follows from (c).

The implication (2) $\Rightarrow$ (1) is obvious in view of (b), (c) and the three spaces theorem (see 12.18.0). ■

The next theorem presents a criterion for verification of whether $\sigma(T)$ is rotation invariant.

13.10. Theorem. *Let X be a Dedekind complete Banach lattice and $T \in dh(X)$. For each $\lambda \in \sigma(T)$ the following conditions* (1) *and* (2) *are equivalent.*

(1) *The circle $\lambda\Gamma \not\subset \sigma(T)$, in particular, spectrum $\sigma(T)$ is not rotation invariant.*

(2) *There are a number $m \in \mathbf{N}$ and a clopen φ_T-invariant subset K_2 of $F_T^{(m)}$ satisfying the next three properties:*

> (a) *K_2 is a retract of $Q(X)$.*
>
> (b) *$\sigma(T|X_0) \cap \lambda\Gamma = \emptyset$, where X_0 is the same as in Theorem* 3.9.
>
> (c) *The quotient space X/X_0 is Dedekind complete, $\dot{T}^{m!} \in Z(X/X_0) = C(K_2)$ and $\lambda \in \sigma(\dot{T})$ though $\lambda\Gamma \not\subset \sigma(\dot{T})$.*

Proof. The implication (2) $\Longrightarrow$ (1) follows immediately from the three spaces theorem formulated in 12.18.0. To prove the converse implication we assume for simplicity that $\lambda = 1$. Let us define numbers $\delta > 0$, $m \in \mathbf{N}$ and an ideal X_δ exactly as it was done in the proof of Theorem 13.4. The equality (13.2) shows that the intersection $\sigma(T|X_\delta) \cap \{\beta : 1 - 2\delta < |\beta| < 1 + 2\delta\}$ is a clopen subset of $\sigma(T|X_\delta)$. Let V be the spectral subspace of X_δ corresponding to this subset, and P_V be the spectral projection from X_δ onto V. By Lemma 13.2.1, $P_V \in dh(X_\delta)$. Let $\varphi = \varphi_{P_V}$ be the mapping of $Q_\delta := supp(X_\delta)$ into Q_δ. Then the definition of V implies that $\varphi(Q_\delta) \subseteq F_T^{(m)}$. Finally letting $K_2 := \varphi(Q_\delta)$, we see that K_2 is a retract of Q_δ, and consequently a retract of $Q(X)$. Moreover, repeating the arguments used in the proof of Theorem 13.4 we can show that $K_2 = supp(V) \cap F_T^{(m)}$, whence K_2 is a clopen subset of $F_T^{(m)}$. The remaining properties (b) and (c) can be verified in the same way as it was done in the proof of Theorem 13.9. ■

13.11. In connection with the previous two theorems several comments are in order. In these theorems we tried to find topological conditions on the mapping φ_T responsible for some symmetry of the spectrum $\sigma(T)$. The non-trivial part of these theorems asserted the existence of a retract K_2 with some nice properties. Also, as we will show in Theorem 13.13, the retract K_2 has a base of φ_T-invariant clopen neighborhoods. Moreover, if the norm on X satisfies (C), then a stronger result is true. Specifically, we can find then a clopen neighborhood V of K_2 such that $\varphi_T^n(V) = K_2$ for some $n \in \mathbf{N}$ (*). When $\sigma(T)$ is not rotation invariant this was verified in Theorem 12.14, and for a non-connected $\sigma(T)$ this may be verified in a similar manner. Actually, for Dedekind complete Banach lattices with (C), the condition (*) is necessary in the following (similar to Theorem 12.14) sense: If (*) holds, then there exists an operator $T_1 \in Z(X)$ such that $\sigma(T_1 T)$ is not connected (respectively, not rotation invariant). It would be interesting to find topological conditions on φ_T which are necessary in the above sense for the class of all Dedekind complete Banach lattices with ($\star$), and for the class of all Banach lattices.

As an application of the previous results we will show next that for a d-homomorphism its spectrum often coincides with its o-spectrum. We again will be using the notations introduced in Theorem 13.1 of the spectral subspaces X_i, X_e, the band $Y = (X_i)^d$, and projections P_e, P_Y.

13.12. Theorem. *Let X be a Dedekind complete Banach lattice, $T \in dh(X)$ and $\lambda \notin \sigma(T)$. If any of the next three conditions is fulfilled*

(a) *X satisfies ($\star$) and T is order continuous, or*

(b) *X satisfies* (C), *or*

(c) $T^* \in dh(X^*)$,

then the resolvent $(\lambda I - T)^{-1}$ *is a regular operator on* X. *In other words, the spectrum* $\sigma(T)$ *coincides with the order-spectrum* $\sigma_o(T)$, *that is,* T *has the same spectrum in the algebra* $L(X)$ *of all continuous operators and in the algebra* $L_r(X)$ *of all regular operators.*

Proof. If $T \in dh(X)$ is invertible, then, as was mentioned in 4.5.1, the inverse T^{-1} also belongs to $dh(X)$. This implies the conclusion of the theorem for $\lambda = 0$, since then $(\lambda I - T)^{-1} = T^{-1}$. So, from now on $\lambda \neq 0$.

Case I. Assume first that no point on the circle $\lambda\Gamma$ belongs to the spectrum, i.e., $\lambda\Gamma \cap \sigma(T) = \emptyset$. In this case $\sigma(T)$ is the union of two disjoint clopen sets σ_i and σ_e, where $\sigma_i \subset D_{|\lambda|}$ and $\sigma_e \cap D_{|\lambda|} = \emptyset$. Let $S = (T|X_e)^{-1}P_e$. (Notice that if σ_i (resp. σ_e) is empty then P_i (resp. P_e) is, of course, the zero operator.) Then as is well known [DS1]

$$(\lambda I - T)^{-1} = \sum_{n=0}^{\infty} \frac{1}{\lambda^{n+1}} T^n P_i - \sum_{n=0}^{\infty} \lambda^n S^{n+1},$$

where the series converges absolutely. By Lemma 13.2.1 the operator $S \in dh(X)$, and consequently $\big\| |S^n| \big\| = \|S^n\|$ for each $n \in \mathbf{N}$. Therefore the series $\sum_{n=0}^{\infty} |\lambda|^n |S^{n+1}|$ converges absolutely and $\sum_{n=0}^{\infty} \lambda^n S^{n+1}$ converges to a regular operator.

To prove that the first series $\sum_{n=0}^{\infty} \frac{1}{\lambda^{n+1}} T^n P_i$ also converges to a regular operator, we will consider conditions (a), (b) and (c) separately.

If (a) holds, then Theorem 13.1 is applicable and as shown in its proof $P_Y P_i = 0$, whence $P_i = (I - P_Y)P_i$. This implies that

$$|P_i| \leq |I - P_Y| \cdot |P_i| = (I - P_Y) \cdot |P_i| = |P_i| - P_Y |P_i|,$$

and hence $P_Y |P_i| = 0$ and $|P_i| X \subset X_i$. Consequently for each $n \in \mathbf{N}$ we have

$$\big\| |T^n P_i| \big\| \leq \big\| |T^n| \cdot |P_i| \big\| \leq \big\| |P_i| \big\| \cdot \|T^n | X_i\|$$

and the series $\sum_{n=0}^{\infty} \frac{1}{|\lambda|^{n+1}} |T^n P_i|$ converges absolutely as $r(T|X_i) < |\lambda|$.

Let (b) holds, and let K_2 be the set introduced in Theorem 13.9. In view of the comments in 13.11, we know that there exists some $m \in \mathbf{N}$ such that

$$\chi_n T^n P_i = 0, \quad n = m, m+1, \dots,$$

where χ_n is the characteristic function of the set $int(\varphi_T^{-n}(K_2))$. Therefore for each $n \geq m$ we have $\chi_n |T^n P_i| = 0$, $|T^n P_i| X \subset X_i$, and the series $\sum_{n=0}^{\infty} \frac{1}{|\lambda|^{n+1}} |T^n P_i|$ converges absolutely.

If (c) holds then by Hart's theorem [Har] (cited in the proof of Proposition 12.20) the mapping $\varphi_{T^{**}}$ is a homeomorphism. Therefore, as the proof of Theorem 13.1 shows, we have $P_i^{**} \in dh(X^{**})$, whence $P_i \in dh(X)$ and $\|T^n P_i\| = \||T^n P_i|\|$. This again implies that the series $\sum_{n=0}^{\infty} \frac{1}{|\lambda|^{n+1}} |T^n P_i|$ converges absolutely.

Case II. Consider the remaining case when $\lambda\Gamma \cap \sigma(T) \neq \emptyset$. This means that condition (1) of Theorem 3.10 is satisfied. Let K_2 be a clopen subset of $F_T^{(m)}$ which exists by condition (2) of the same theorem. We denote by $K_2^{\#}$ the smallest closed φ_T^{-1}-invariant subset of $Q(X)$ containing K_2. The proof of Theorem A.12 shows that the set $K_2^{\#}$ is clopen in $Q(X)$. Let h denotes its characteristic function. Using this function we introduce $X_1 = hX$ and $X_2 = (\mathbf{1} - h)X$, two complementary bands in X.

Clearly $Th = hT$; consequently the spaces X_1 and X_2 are T-invariant and

$$(\lambda I - T)^{-1} = ((\lambda I - T)|X_1)^{-1} h + ((\lambda I - T)|X_2)^{-1}(\mathbf{1} - h).$$

Let $R_1 = ((\lambda I - T)|X_1)^{-1}$ and $R_2 = ((\lambda I - T)|X_2)^{-1}$. To prove the theorem it will suffice to show that the operators R_1 and R_2 are regular. The regularity of R_2 follows directly from Case I, as $\sigma(T|X_2) \cap \lambda\Gamma = \emptyset$.

Consider now the operator $T|X_1$, and recall that by $\dot{T}$ we denoted in Theorem 3.10 the quotient operator induced by T on X/X_0. The definition of the set K_2 in the proof of Theorem 13.10 shows that $\sigma(T|X_1)$ consists of two clopen parts, one of which is a closed disk centered at zero, and another coincides with $\sigma(\dot{T})$. Imitating the notation introduced in Theorem 3.1 we denote by X_{1i} (resp. X_{1e}) the spectral subspace of X_1 corresponding to the internal (resp. external) part of the spectrum $\sigma(T|X_1)$, and by P_{1i} (resp. P_{1e}) the corresponding spectral projection. Then

$$R_1 = \sum_{n=0}^{\infty} \frac{1}{\lambda^{n+1}} T^n P_{1i} + ((\lambda I - T)|X_{1e})^{-1} P_{1e}.$$

The operator $\sum_{n=0}^{\infty} \frac{1}{\lambda^{n+1}} T^n P_{1i}$ is regular in view of Case I.

It remains only to verify that the operator $R_e := ((\lambda I - T)|X_{1e})^{-1} P_{1e}$ is regular. Let J be the restriction to X_{1e} of the canonical quotient map $X \mapsto X/X_0$. Repeating the arguments from the proof of Theorem 13.9 we can show that J is an isomorphism from X_{1e} onto X/X_0 and that $T|X_{1e} = J^{-1}\dot{T}J$. By Theorem 13.10.2(c) we know that $(\dot{T})^{m!} \in Z(X/X_0)$, and hence $(\lambda^{m!}I - \dot{T}^{m!})^{-1} \in Z(X/X_0)$. Also we will need the following elementary identity

$$(\lambda I - \dot{T})^{-1} = (\lambda^{m!-1}I + \lambda^{m!-2}\dot{T} + \cdots + \dot{T}^{m!-1})(\lambda^{m!}I - \dot{T}^{m!})^{-1}.$$

This identity and the definition of the operator R_e imply that

$$R_e = J^{-1}(\lambda^{m!-1}I + \lambda^{m!-2}\dot{T} + \cdots + \dot{T}^{m!-1})(\lambda^{m!}I - \dot{T}^{m!})^{-1}JP_{1e}.$$

The regularity of R_e will follow if we show that for each $k \leq m! - 1$ the operator

$$R_{ek} = J^{-1}\dot{T}^k(\lambda^{m!}I - \dot{T}^{m!})^{-1}JP_{1e}$$

is regular. Moreover, we will see that each $R_{ek} \in dh(X_1)$. To show this, we begin with a special case when X_1 satisfies (C) and T is order continuous. Indeed, it follows from Theorem 13.1 that in this case K_2 is a clopen subset of $Q(X)$ and we can identify X/X_0 with the band $\chi_{K_2}X$. Then $Jx = \chi_{K_2}x$ for each $x \in X_{1e}$. Since the mapping φ_T has period $m!$ on K_2, the operator $\dot{T}$ can be identified with a periodic d-automorphism of the band $\chi_{K_2}X$. Now the desired conclusion that $R_{ek} \in dh(X_1)$ follows from the fact that two elements $x', x'' \in X_{1e}$ are disjoint in X if and only if $Jx'\mathrm{d}Jx''$.

The general case of arbitrary T and X_1 can be reduced to the preceding one by considering R_{ek}^{**}. ■

Recall that a regular operator T on a Banach lattice X is said to be band irreducible if the only T-invariant bands in X are the trivial ones, that is, X itself and $\{0\}$. One of the classical results regarding these operators is the Andô-Krieger Theorem stating that each positive band irreducible kernel operator has a positive spectral radius. We refer to [Sch] and [Zaa] for the basic theory about band irreducible operators and to

[AAB] for the latest developments related to the Andô-Krieger Theorem. As our next theorem shows in the case of band irreducible d-homomorphisms we can completely describe their spectrum.

13.13. Theorem. *Let X be an infinite-dimensional Dedekind complete Banach lattice and $T \in dh(X)$ be band irreducible. Then the spectrum $\sigma(T)$ is a disc or an annulus.*

Proof. We begin by showing that $\sigma(T)$ is rotation invariant. If not, then by Remark 12.12.1 the set $F_T^{(m)} \neq \emptyset$ for some m. Let $t \in F_T^{(m)}$ and $p \in \mathbf{N}$ be the period of this point. Consider the orbit $O_t = \{t, \varphi_T(t), \cdots, \varphi_T^{p-1}(t)\}$ of this point. As follows easily from the Frolik invariant base theorem (see Corollary A.4.1) there is a base of clopen φ_T-invariant neighborhoods of this orbit. Because of this and since $\dim X = \infty$, there exists a non-void clopen subset $V \subset E_T$ such that $\varphi_T(V) \subset V$ and $E_T \setminus V \neq \emptyset$. Then $Y = \chi_{E_T \setminus V} X$ is a non-trivial T-invariant band in X, in spite of the condition that T is band irreducible. Thus $\sigma(T)$ is rotation invariant.

To prove that $\sigma(T)$ is a disc or annulus it remains to prove that $\sigma(T)$ is connected. Assume, contrary to what we claim, that $\sigma(T)$ is not connected. Being rotation invariant and not being connected the set $\sigma(T)$ satisfies the condition (1) of Theorem 13.9. Therefore it satisfies the condition (2) too, and hence there exists a subset K_2 of $Q(X)$ satisfying the properties (a) through (e) of that condition (2).

If we verify now that the set K_2 has a base of φ_T-invariant clopen neighborhoods, then we arrive at a contradiction to our assumption that T is band irreducible. To this end let V be an arbitrary clopen neighborhood of K_2. Replacing V by $V \cap E_T$ we can assume also that $V \subseteq E_T$. Consider the clopen set $A = V \cap \varphi^{-1}(Q(X) \setminus V)$. If $A = \emptyset$, then the set V is itself φ_T-invariant, so assume that $A \neq \emptyset$. We will denote by $A^{\#}$ the smallest closed φ_T^{-1}-invariant subset containing A. A rather detailed theory of the properties of the set $A^{\#}$ is presented at the beginning of Appendix A. Here we need only a simple fact that $A^{\#}$ is clopen, which can be verified in a straightforward manner.

Consider the operator $T_V = \chi_V T$. Obviously the set E_{T_V} (the support of the mapping φ_{T_V}) coincides with V, and on this set the mapping φ_{T_V} coincides with φ_T. Therefore, the ideal X_0 introduced in Theorem 13.9(b) is T_V-invariant, the quotient operator $\dot{T}_V$ on X/X_0 coincides with $\dot{T}$, and $\sigma(\dot{T}_V)$ does not intersect the disc D_r.

Notice next that we have the following equality

$$(T_V)^n = \chi_V \cdot (\chi_V \circ \varphi_T) \cdot (\chi_V \circ \varphi_{T^2}) \cdots (\chi_V \circ \varphi_{T^{n-1}}) T^n, \; n \in \mathbf{N}, \tag{13.7}$$

which implies that $r(T_V|X_0) \le r(T|X_0) < r$. The three spaces theorem implies now that $\sigma(T_V)$ satisfies the condition (1) of Theorem 13.9. Let Y_V be the spectral subspace of the operator T_V corresponding to the set $\sigma(T_V) \setminus D_r$. The operator $T_V|Y_V$ is invertible and from (13.7) it follows that $supp(Y_V) \cap A^{\#} = \emptyset$. Since $K_2 \subseteq supp(Y_V)$, the set $W = V \setminus A^{\#}$ is a clopen neighborhood of K_2 contained in V. The very definition of $A^{\#}$ implies that W is φ_T-invariant, and we are done. ■

We conclude this section by considering the case when the conjugate operator T^* is also a d-homomorphism.

13.14. Theorem. *Let X be a Banach lattice and $T \in dh(X)$ be such that $T^* \in dh(X^*)$. If $T^m \delta\, T^n$ for any distinct $m, n \in \mathbf{N} \cup \{0\}$, then $\sigma(T)$ is rotation invariant.*

Proof. If $\sigma(T)$ is not rotation invariant, then it was shown in the course of the proof of Theorem 13.4 (without using condition ($\star$)) that there is a non-zero subspace V of X such that for some $m \in \mathbf{N}$ the set $supp(jV) \subset \mathrm{cl}\bigcup_{k=1}^{\infty} \varphi_T^{-k}(F_{T^{**}}^{(m)})$ (See also Remark 13.2). However, since T^{**} is order continuous and $T^* \in dh(X^*)$, the mapping $\varphi_{T^{**}}$ is one-to-one by Hart's theorem [Har], and therefore clearly, $supp(jV) \subset F_{T^{**}}^{(m)}$. In view of Proposition 11.36 this contradicts the independence of powers of T. ■

13.15. Proposition. *Let X be a Dedekind complete Banach lattice and $T \in dh(X)$ be such that $T^* \in dh(X^*)$. Then $\sigma^{(m)} \subseteq \sigma(T)$ and $\sigma(T) \setminus \cup_{m=1}^{\infty} \sigma^{(m)}$ is rotation invariant.*

Proof. Remark 11.37 and the fact that $\varphi_{T^{**}}$ is a homeomorphism clearly imply that $\varphi_T^{-1}(G_T^{(m)}) = G_T^{(m)}$. Therefore $g^{(m)}T = Tg^{(m)}$, and consequently the set $\sigma^{(m)}(=$

$\sigma(g^{(m)}T|g^{(m)}X)) \subseteq \sigma(T)$. Moreover, $\sigma^{(m)} = \sigma^{(m)}(T^{**})$. As a result we have that $\sigma(T)\setminus \cup_m \sigma^{(m)} = \sigma(T^{**})\setminus \cup_m \sigma^{(m)}(T^{**})$, and the set on the right hand side is rotation invariant by Proposition 12.18. ■

13.16. Corollary. *Assume the conditions of the previous proposition and that* $\sigma(T) \subset \{re^{i(\theta+\theta_0)} : 0 \leq r < \infty, 0 \leq \theta < 2\pi(1 - \frac{1}{m+1})\}$ *for some* $\theta_0 \in \mathbf{R}$ *and* $m \in \mathbf{N}$. *Then* X *is a direct sum of two disjoint* T*-invariant bands* X_1 *and* X_2 *such that* $\sigma(T|X_1) = \{0\}$ *and* $T^p|X_2 \in Z(X_2)$, *where* p *is the least common multiple of numbers* $1, 2, \ldots, m$.

14. Examples

The spaces of continuous functions

14.1. Let $X = C(K)$ be the Banach lattice of all continuous complex valued functions on a compact Hausdorff space K. It is clear that X is also an operator $C(K)$-module and the structure of this module coincides with the original structure of the Banach lattice X. The center $Z(X)$ of this module coincides with $C(K)$.

Take an arbitrary $x \in X$. It is easy to see that the ideal $X(x)$ consists of all continuous functions which are equal to zero outside the support set of the function x, and the center $Z(X(x))$ of this ideal coincides with the algebra of all bounded continuous functions on the open set $\{t \in K : x(t) \neq 0\}$.

Let K_1 and K_2 be two compact Hausdorff spaces, $X = C(K_1)$, $Y = C(K_2)$ and $T \in dh(X, Y)$. We consider $\mathcal{O}_T := \{t \in K_2 : T\mathbf{1}(t) \neq 0\}$ which is an open subset of K_2. By Theorem 6.7 there exists a unique homomorphism $\gamma : Z(X) \to Z(Y_{TX})$, such that $T(f \cdot x) = \gamma(f) \cdot Tx$ for all $x, f \in C(K_1)$. Since $\mathcal{O}_T$ is homeomorphic to a dense open subset of the maximal ideal space of the algebra $Z(Y_{TX})$, there exists the unique continuous mapping ψ_T from $\mathcal{O}_T$ to K_1 corresponding to this algebraic homomorphism γ, i.e., such that $\gamma(f) = f \circ \psi_T$.

14.1.1. If K_2 is extremally disconnected, then this mapping ψ_T coincides with the mapping φ_T which was used throughout Part 1.

14.2. Proposition. *For $S, T \in dh(X, Y)$ the following conditions are equivalent.*

(1) $S \bar{\delta} T$.

(2) *The open set $\mathcal{O} = \mathcal{O}_S \cap \mathcal{O}_T$ is not empty and the mappings ψ_S and ψ_T coincide on it.*

Proof. (1) $\Rightarrow$ (2). By Lemma 11.4(3) there exist $g_1, g_2 \in Z(Y_{TX})$ such that $g_1 S = g_2 T \neq 0$. This obviously implies that $\psi_S = \psi_T$ on the non-empty open set

$$\{t \in K_2 : (g_1 S\mathbf{1})(t) = (g_2 T\mathbf{1})(t) \neq 0\}.$$

(2) $\Rightarrow$ (1). Pick an arbitrary $g_1 \in C(K_2)$ such that $supp(g_1) \subset \mathcal{O}$. Then $(S\mathbf{1})(t) \neq 0$ at each point $t \in supp(g_1)$ and $g_1 T = g_2 S \neq 0$, where $g_2(t) = (g_1 T\mathbf{1})(t)/(S\mathbf{1})(t)$ for $t \in supp(g_1)$ and $g_2(t) = 0$ for $t \in K_2 \backslash supp(g_1)$. By Lemma 11.4 $S\,\bar{\delta}\,T$. ■

Let us consider a special case of the previous situation when $X = Y$, i.e., $X = C(K)$ and $T \in dh(T)$. If $0 \leq m < n$, then clearly $\mathcal{O}_{T^n} \subset \mathcal{O}_{T^m}$ and $\mathcal{O}_{T^n}$ consists precisely of all those points t in $\mathcal{O}_T$ for which $\psi_T(t) \in \mathcal{O}_T$, $\psi_T^2(t) = \psi_T(\psi_T(t)) \in \mathcal{O}_T, \ldots, \psi_T^{n-1}(t) \in \mathcal{O}_T$. The mapping ψ_{T^n} is defined on $\mathcal{O}_{T^n}$ and on this set $\psi_{T^n} = \psi_T^n$. Let us introduce

$$F_T^{(n)} = \{t \in \mathcal{O}_{T^n} : \psi^l(t) = t \;\text{ for some } l \leq n\}.$$

Then the previous proposition implies the following corollary.

14.3. Corollary. *For any $T \in dh(X)$, where $X = C(K)$, the following statements are equivalent.*

(1) *$T^m\,\bar{\delta}\,T^n$ for all $0 \leq m < n$.*

(2) *For arbitrary $p, n \in \mathbf{N}$ the set $\operatorname{int}(\psi_T^{-p}(F_T^{(n)}))$ is empty.*

14.4. Remark. As we know, a d-homomorphism $T \in dh(X)$ is order continuous if and only if the mapping φ_T is open. Therefore, in this case $T^m\,\bar{\delta}\,T^n$ for all $0 \leq m < n$ if and only if $I\,\bar{\delta}\,T^m$ for all $m \in \mathbf{N}$ if and only if $\operatorname{int}(F_T^{(m)})) = \emptyset$ for all $m \in \mathbf{N}$.

14.5. Remark. An operator $T \in dh(X)$ is nice, i.e., $T \in dh_n(X)$ if $supp(T\mathbf{1})$ is a clopen subset of K. In this case the set $\mathcal{O}_T$ coincides with the set E_T introduced in 11.11.1 and the mapping ψ_T coincides with φ_T.

Proposition 14.2 and Theorem 12.11 imply now the following corollary.

14.6. Corollary. *Let $T \in dh(X)$ satisfies either of the conditions of Corollary* 14.3. *Then the spectrum $\sigma(T)$ is rotation invariant.*

14.7. Corollary. *Let $T \in dh(X)$ be order continuous and let $\operatorname{int}(F_T^{(m)}) = \emptyset$ for all $m \in \mathbf{N}$. Then the spectrum $\sigma(T)$ is rotation invariant.*

14.8. Remark. We refer to [Kit3] for a more detailed investigation of the spectrum of the nice d-endomorphisms on $C(K)$-spaces. As an example illustrating the previous corollary consider a domain Ω in $\mathbf{C}^n$ and an analytical transformation ψ of Ω onto Ω, which can be continuously extended to $K = \mathrm{cl}(\Omega)$. Then the operator T on $C(K)$ defined by $Tf = a(f \circ \psi)$ has a rotation invariant spectrum for each $a \in C(K)$.

14.9. Remark. Any d-endomorphism on the space $C(K)$ admits the following representation.

$$(Tf)(t) = \begin{cases} (T\mathbf{1})(t)f(\psi_T(t)), & t \in \mathcal{O}_T \\ 0, & t \in K \backslash \mathcal{O}_T \end{cases}$$

Banach lattices with quasi-interior points

14.10. Let $X = X(x_0)$ and $Y = Y(y_0)$ be two Banach lattices with quasi-interior points x_0 and y_0, respectively. Let $S, T \in dh(X,Y)$ and let $v = y_0 + |Sx_0| + |Tx_0|$. Obviously $Y = Y(v)$ and in view of Theorem 3.3 both operators S and T map the ideal X_{x_0} into Y_v. Therefore, by Lemma 11.4(3), we have $S\,\delta\,T$ if and only if S and T are independent as operators from X_{x_0} into Y_v. But X_{x_0} is order isomorphic to $Z(X) = C(K_1)$ and Y_v is order isomorphic to $Z(Y) = C(K_2)$ and, thus Proposition 14.2 is applicable.

14.11. Remark. The sets $\mathcal{O}_S = \{t \in K_2 : S\mathbf{1}(t) \neq 0\}$, $\mathcal{O}_T = \{t \in K_2 : T\mathbf{1}(t) \neq 0\}$, and the mappings $\psi_S : \mathcal{O}_S \to K_1$, $\psi_T : \mathcal{O}_T \to K_1$ are essentially independent of the choice of quasi-interior points in X and Y in the sense that if we take another pair of quasi-interior points in X and Y, then the corresponding mappings will coincide with ψ_S and ψ_T respectively on open dense subsets of $\mathcal{O}_S$ and $\mathcal{O}_T$.

14.12. Remark. Let $X = X(x_0)$ be a Banach lattice with a quasi-interior point x_0 and let $T \in dh(X)$. Introduce $u = x_0 + \sum_{n=1}^{\infty} \varepsilon^n Q_n(x_0)$, where $Q_1(x_0) = (|T_1| + |T_2|)(x_0)$, $Q_2(x_0) = (|T_1^2| + |T_1T_2| + |T_2T_1| + |T_2^2|)(x_0)$, etc, and let $0 < \varepsilon < (2\max(\|T_1\|, \|T_2)\|)^{-1}$. Then $TX_u \subset X_u$, and therefore we can apply Corollary 14.3.

14.13. Remark. A d-endomorphism $T \in dh(X)$ is order continuous if and only if $\psi_T : \mathcal{O}_T \to K$ is open, and thus Remark 14.4 is applicable.

Now applying Theorem 13.4 we obtain the following proposition.

14.14. Proposition. *Let X be a Banach lattice with a quasi-interior point and let X satisfies the condition $(\star)$ introduced at the very beginning of Section 13. Assume also that an operator $T \in dh(X)$ satisfies either of the conditions of Corollary 14.3. Then the spectrum $\sigma(T)$ is rotation invariant.*

14.15. Proposition. *Let the assumptions of the previous proposition hold and let T be order continuous and $\mathrm{int}(F_T^{(m)}) = \emptyset$ for all $m \in \mathbf{N}$. Then the spectrum $\sigma(T)$ is rotation invariant.*

14.15.1. Example. Let (Ω, Σ, μ) be a σ-finite measure space, and let ψ be a measurable transformation of Ω. Assume that for a measurable function e the operator $Tx = e(x \circ \psi)$ is defined and bounded on $L^p(\mu)$ for some $p \in [1, \infty)$. If the set of all ψ-periodic points is of measure zero, then $\sigma(T)$ is rotation invariant. Moreover, if ψ is ergodic (i.e., $\psi(A) \subseteq A \Rightarrow (\mu(A) = 0 \text{ or } \mu(\Omega \backslash A) = 0)$), then $\sigma(T)$ is either a disk or an annulus.

14.16. Remark. As follows from Proposition 13.5, the assumption $(\star)$ in Propositions 14.14 and 14.15 is essential.

14.17. Remark. Propositions 14.14 and 14.15 imply as special cases the results of [Pet], [Lat] and [AL].

The spaces of continuous vector-valued functions

14.18. Let K_1, K_2 be two compact Hausdorff spaces and G_1, G_2 be two Banach spaces. We denote by $X_j = C(K_j, G_j)$, $(j = 1, 2)$, the Banach space of all G_j-valued continuous functions f on K_j with the standard supremum norm

$$\|f\|_j = \sup_{t \in K_j} \|f(t)\|_{G_j}.$$

It is easy to see that X_j is a $C(K_j)$-module.

Let $T \in dh(X_1, X_2)$. For each $x \in X_1$ the center of the lattice $X_2(Tx)$ coincides with the space of all bounded continuous functions on the open set $\mathcal{O}_x = \{t \in K_2 : (Tx)(t) \neq 0\}$. Therefore, by Theorem 6.7, the operator T generates a continuous mapping $\psi_{T,x} : \mathcal{O}_x \to K_1$. It is not difficult to verify that if $\mathcal{O} = \mathcal{O}_x \cap \mathcal{O}_y \neq \emptyset$ for some $x, y \in X_1$, then the mappings $\psi_{T,x}$ and $\psi_{T,y}$ coincide on $\mathcal{O}$. Therefore there exists a mapping $\psi_T : \mathcal{O}_T \to K_1$, where $\mathcal{O}_T := \cup_{x \in X_1} \mathcal{O}_x$. Using these notations we have the following proposition.

14.19. Proposition. *Let $S, T \in dh(X, Y)$ and $S \bar{\delta} T$. Then the mappings ψ_S and ψ_T coincide on a non-empty open subset of $\mathcal{O}_S \cap \mathcal{O}_T$.*

As the following example shows, the converse statement is not true in general.

14.20. Example. Let $X_1 = X_2 = C([0,1], \mathbf{C}^2)$ and let

$$Sx(t) := \begin{bmatrix} 1 & 0 \\ 0 & 0 \end{bmatrix} x(t), \quad Tx(t) := \begin{bmatrix} 0 & 0 \\ 0 & 1 \end{bmatrix} x(t).$$

Definition 11.8 easily implies that $S \, \delta \, T$, though $\psi_S(t) = \psi_T(t) = t$ for all $t \in [0,1]$. Nevertheless for d-endomorphisms the following result holds.

14.21. Proposition. *Let $X = (C(K), G)$, $T \in dh(X)$ and $0 \leq m < n$. Then $T^m \, \bar{\delta} \, T^n$ if and only if $\psi_{T^m} = \psi_{T^n}$ on some non-empty open subset $\mathcal{O}$ of the set $\mathcal{O}_{T^n}$.*

Proof. Assume that $\psi_{T^m} = \psi_{T^n}$ on some non-empty open subset $\mathcal{O} \subset \mathcal{O}_{T^n}$. Take $f \in C(K)$ and $x \in X$ such that $supp(f) \subset \mathcal{O}$ and $fT^n x \neq 0$. Then $f(T^m)^{x_1} = f(T^n)^x$, where $x_1 = T^{n-m}x$. ■

14.22. Remark. If $m < n$, $m, n \in \mathbf{N}$, then the following inclusions are true $\mathcal{O}_{T^n} \subset \psi_T^{-(n-m)}(\mathcal{O}_{T^m}) \subset \mathcal{O}_{T^m}$ but, in general, $\mathcal{O}_{T^n} \neq \psi_T^{-(n-m)}(\mathcal{O}_{T^m})$. To see this take, for instance, the operator T defined in Example 14.20. Then we have $\mathcal{O}_T = \psi_T^{-1}(\mathcal{O}_T) = [0,1]$, but $T^2 = 0$ and $\mathcal{O}_{T^2} = \emptyset$.

We leave it to the reader to formulate the analogs of Corollaries 14.6 and 14.7 for d-endomorphisms on the space $C(K, G)$.

Lattice normed spaces

14.23. Recall the definition of a lattice normed space. A triple (X, p, G), where X is a linear space, G is a Dedekind complete Banach lattice and p is a mapping from

X into G_+, is said to be a **lattice normed space** provided the following conditions are satisfied:

(1) $p(x) = 0$ iff $x = 0$,

(2) $p(x_1 + x_2) \leq p(x_1) + p(x_2)$ for all $x_1, x_2 \in X$,

(3) $p(\lambda x) = |\lambda| p(x)$ for all $x \in X$ and $\lambda \in \mathbf{C}$.

(4) if for some $x \in X$ the element $p(x)$ is the sum of two disjoint elements in G, i.e., $p(x) = e_1 + e_2$ and $e_1 \wedge e_2 = 0$, then there exist $x_j \in X$ such that $x = x_1 + x_2$ and $p(x_j) = e_j$ for $j = 1, 2$.

The conditions (1)–(3) are the usual conditions on the vector norm and (4) is referred to as Kantorovich's splitting property. For details on lattice normed spaces we refer to [KS] and [KVP].

It is immediate that by letting $\|x\| = \|p(x)\|_G$ we turn X into a normed space. If we denote the completion of X with respect to this norm by $\hat{X}$, then the triple $(\hat{X}, p, G)$ is also a lattice normed space. This allows us to assume from the very beginning that X is norm-complete. Obviously, $p : X \to G$ is continuous.

Let P denote any band projection in G. It is easy to see that there exists a contractive projection h_P of X onto the subspace $p^{-1}(PG)$ such that $ker(h_P) = p^{-1}((I-P)G)$. This mapping $h : P \mapsto h_P$ generates an isometric representation of $Z(G) = C(Q(G))$ in $L(X)$, thus turning X into an operator $Z(G)$-module. Splitting property (4) guarantees that $supp_Q(x) = supp_Q(p(x))$ for all $x \in X$, where, as usual, $Q = Q(G)$ denotes the Stone space of G. Thus, X is an order complete $C(Q(G))$-module and by Proposition 11.20(2) we have that $Z(X) = Z(G)$.

Let X be a lattice normed space and $x \in X$. It is easy to see that the ideal $X(x)$ is order isometric to the ideal $G(p(x))$. Hence, if the Banach lattice G satisfies (C), then our module X likewise satisfies (C) (see Definition 12.10.1), and consequently we get the following corollary.

14.24. Corollary. *Let (X, p, G) be a lattice normed space, where G satisfies* (C). *If $T \in dh(X)$ and $T^m \,\delta\, T^n$ for all $0 \leq m < n$, then the spectrum $\sigma(T)$ is rotation invariant.*

Recall that in Proposition 11.26(2) we have presented a condition which is equivalent to the independence of operators and which is given in terms of the mappings of the underlying compact space.

14.25. Example. Let G be a Banach function space on a measure space (Ω, μ), and Y be a Banach space. We denote by X the space of all strongly measurable Y-valued functions $x(\cdot)$ on Ω, such that the function $p(x) = \|x(\cdot)\|_Y \in G$. Then (X, p, G) is a lattice normed space.

Let M be a strongly measurable $L(X)$-valued function on Ω, ψ be a measurable mapping of Ω such that $\mu(\psi(A)) > 0$ whenever $\mu(A) > 0$, and T be a weighted composition operator generated by M and ψ, i.e., $Tx = M(x \circ \psi)$. If T is bounded on X, then T belongs to $dh(X)$ and is order continuous. So we can apply Proposition 12.15 and Remark 12.16 to describe its spectral properties. In particular, $T^m \,\delta\, T^n$ for all $0 \leq m < n$ if and only if $\mu(F_n) = 0$, $n \in \mathbf{N}$, where F_n is the set of all fixed points of the mapping ψ^n. If this is the case, then $\sigma(T)$ is rotation invariant.

Tensor products of V. Levin

14.26. Let G be a Banach lattice and Y be a Banach space. In the algebraic tensor product $Y \otimes G$ we introduce the following cross-norm

$$\|z\| = \inf \|\sum_{k=1}^{n} \|y_k\|_Y \cdot |e_k|\,\|_G,$$

where the infimum is taken over all representations of z in the form $z = \sum_{k=1}^{n} y_k \otimes e_k$, where $y_k \in Y$ and $e_k \in G$. This cross-norm was introduced by V. Levin [Lev] and it turned out to be very useful. The completion of the space $Y \otimes G$ with respect to this norm will be denoted by X. It is easy to see that X is a $Z(G)$-module. Let us verify that its center $Z(X)$ coincides with $Z(G)$. Indeed, if $h \in Z(X)$ and a net $\{h_\alpha\}$ converges to h in the strong operator topology on $L(X)$, then the convergence $h_\alpha y \otimes e \to h(y \otimes e)$ for all $e \in G, y \in Y$ implies that there exists a $g \in Z(G)$ such that $h_\alpha \to g$ in the strong operator topology of the space $L(G)$. This implies that $h(y \otimes e) = gy \otimes e$ for all $y \in Y$ and $e \in G$.

It follows from [Lev], Theorem 4.2 and Lemma 4.3, that there exists a vector norm $p : X \to G_+$ satisfying for each $x \in X$ the following two properties:

$$\|x\|_X = \|p(x)\|_G$$

and

$$p(fx) = |f|p(x), \quad \text{for each } f \in Z(X). \tag{$*$}$$

It is plain to see that $(*)$ implies that for each $x \in X$ the lattice $X(x)$ is order-isomorphic and isometric to $G(p(x))$, and consequently the module X satisfies the condition (C) provided the Banach lattice G satisfies it. Therefore, our next result is immediate from Theorem 12.11.

14.27. Corollary. *Let the Banach lattice G satisfies (C). If $T \in dh(X)$ and $T^m \,\delta\, T^n$ for all $0 \le m < n$, then the spectrum $\sigma(T)$ is rotation invariant.*

14.28. In two important special cases we are able to express the condition of the independence of different powers of a d-homomorphism in terms of the mappings (on underlying compact space) generated by these powers.

(1) The lattice G is Dedekind complete. In this case $(*)$ implies that (X, p, G) is a lattice normed space and we can apply Corollary 14.24. Moreover, if G is a Banach function space on a measure space (Ω, μ) and $G(Y)$ denotes the space of all strongly measurable Y-valued functions $y(\cdot)$ on Ω for which the function $\|y(\cdot)\|_Y$ belongs to G, then the space X introduced in 14.26 coincides with the closure in $G(Y)$ of the subspace of all "elementary" functions $\sum_{k=1}^{n} e_k(\cdot)y_k$, where $e_k \in G$ and $y_k \in Y$.

(2) G is a Banach lattice with a quasi-interior point e. Let $T \in dh(X)$. For each $x \in X$ we consider an open subset $\mathcal{O}_x = \{k \in K : (e \wedge p(Tx))(k) > 0\}$ and let $\mathcal{O}_T = \bigcup_{x \in X} \mathcal{O}_x$. From 14.1 and 14.10 it follows that for each $T \in dh(X)$ there exists a unique mapping $\psi_T : \mathcal{O}_T \to K$, and it is easy to verify that this mapping has the following property.

14.29. Proposition. *If $T, S \in dh(X)$ and $T\bar{\delta}S$, then $\psi_T \equiv \psi_S$ on some open non-empty subset of $\mathcal{O}_T \cap \mathcal{O}_S$. Consequently, if the latter condition fails for any two d-homomorphisms, then they are independent.*

14.30. Remark. If $G = C(K)$, where K is a compact Hausdorff space, then X coincides with the space $C(K, Y)$ of all Y-valued continuous functions on K [Lev]. Therefore, three out of the four examples considered in this section (namely, the spaces of continuous functions, the spaces of continuous vector-valued functions, and Banach lattices with quasi-interior point) are special cases of the Levin tensor product depending on a choice of the Banach lattice G and the Banach space Y. Consequently, Corollary 14.27 and Proposition 14.29 contain as special cases the previous results devoted to these particular examples.

Appendices

A. Continuous mappings of extremally disconnected compact spaces

In Sections 11, 12 and 13 we have exploited several times the special properties of continuous mappings on extremally disconnected spaces, in particular, Frolík's theorems and their generalizations obtained in [Kit2]. Since the last work is not readily available, and since the proofs in [Kit2] do not contain many necessary details, we present below the improved and self-contained proofs of those basic results about continuous mappings on extremally disconnected spaces, which we used in the previous sections. Our approach is different from the original approach developed by Frolík, and it is based on the investigation of the backward orbit of the set of fixed points. This allows us not only to reprove Frolík's theorems, but also to generalize them and to obtain several other results of independent interest.

Throughout this section K denotes an arbitrary **extremally disconnected** compact Hausdorff space, $\varphi : K \to K$ is a **continuous mapping**, and $\mathbf{F}$ denotes the **set of all fixed points of** φ. For an arbitrary subset A of K we denote by $A^\#$ the smallest closed φ^{-1}-invariant subset of K containing A, that is, $A \subseteq A^\#$, $\varphi^{-1}(A^\#) \subseteq A^\#$ and $A^\#$ is the smallest closed set satisfying the previous two properties.

The existence of $A^\#$ is obvious, but we also need a constructive description of $A^\#$ which is given next. For each ordinal α, we use transfinite induction to introduce a set A_α as follows:

1. $A_0 = \mathrm{cl}(A)$.
2. If $\alpha > 0$ is a nonlimit ordinal, then we let $A_\alpha = A_{\alpha-1} \cup \varphi^{-1}(A_{\alpha-1})$.
3. If α is a limit ordinal, then we let $A_\alpha = \mathrm{cl}(\bigcup_{\beta<\alpha} A_\beta)$.

Clearly, there exists an ordinal γ such that $\varphi^{-1}(A_\gamma) \subseteq A_\gamma$, and it is plain to see that $A_\gamma = A^\#$.

Observe that if A is clopen, then as our construction shows the set $A^\#$ is likewise clopen. In the next lemma we collect several simple properties of the operation #. The proofs are straightforward and omitted.

A.1. Lemma. *Let A, B be two subsets of K. Then the following statements hold.*

1) $A = A^{\#}$ *if and only if A is closed and* $\varphi^{-1}(A) \subseteq A$.

2) $(A^{\#})^{\#} = A^{\#}$.

3) *For any* $t_0 \in A^{\#} \setminus \bar{A}$ *the point* $\varphi(t_0) \in A^{\#}$.

4) *If* $A^{\#} \cap B^{\#} = \emptyset$, *then* $\varphi(A^{\#}) \cap B^{\#} = \emptyset$.

5) *If A and B are two arbitrary φ-invariant disjoint clopen sets, then* $A^{\#} \cap B^{\#} = \emptyset$.

6) *If* $A = A^{\#}$ *and* $E = \operatorname{int}(A)$, *then* $E = E^{\#}$

In the next lemma $b, c \in \mathbf{C}$ denote arbitrary numbers such that $b \neq \pm c$ and $c \neq 0$. This lemma is crucial for what follows in this section.

A.2. Lemma. *Let A be an arbitrary closed subset of K satisfying $A \cap F = \emptyset$, where F is the set of all fixed points of φ. Then there exists a function $f \in C(A^{\#})$ such that for each $t \in A^{\#}$ the image $f(t) \in \{b, c, -c\}$ and moreover $f(t) \neq f(\varphi(t))$, whenever the point $\varphi(t)$ also belongs to $A^{\#}$.*

Proof. There is nothing to prove if $A = \emptyset$, so assume that $A \neq \emptyset$. Since A and F are closed and disjoint they can be separated by clopen sets, and so, without loss of generality, we can assume that A itself is clopen. The proof will be a two stage induction argument. Since $\varphi(t) \neq t$ for each $t \in A$, the compactness of A implies that for some $n \in \mathbf{N}$ there exist pairwise disjoint clopen subsets $A_1, A_2, \ldots, A_n$ of A such that $A = \bigcup_{j=1}^{n} A_j$ and $\varphi(A_j) \cap A_j = \emptyset$, $j = 1, 2, \ldots, n$.

The first induction will be on n. We begin with $n = 1$. That is, we assume that $A \cap \varphi(A) = \emptyset$, whence clearly, $A \cap \varphi^{-1}(A) = \emptyset$. To construct f at this stage we will use induction on α. First we define f on the set A_1 by letting $f \equiv b$ on A and by letting $f \equiv c$ on $\varphi^{-1}(A)$. Let now α be some ordinal, and suppose that f has been already defined on A_β for each $\beta < \alpha$. If α is a nonlimit ordinal, then by definition of A_α we have

$$A_\alpha \backslash A_{\alpha-1} = \varphi^{-1}(A_{\alpha-1}) \backslash A_{\alpha-1}$$

and we extend f to $A_\alpha \backslash A_{\alpha-1}$ by setting

$$f(t) = -f((\varphi(t)), \quad t \in A_\alpha \backslash A_{\alpha-1}.$$

If α is a limit ordinal, then, by the induction hypothesis, f is defined on the open set $\bigcup_{\beta<\alpha} A_\alpha$, and consequently (because K is extremally disconnected) f can be continuously extended to $A_\alpha = \mathrm{cl}(\bigcup\limits_{\beta<\alpha} A_\beta)$. Repeating this process we see that f can be defined continuously on all of the set $A^\#$. By the very definition, f takes on only the values c and $-c$ on $A^\# \setminus A$.

It remains to verify that $f(t) \neq f(\varphi(t))$, whenever t and $\varphi(t) \in A^\#$. The following three possible cases prove the claim.

i) The point $t \in A$ but $\varphi(t) \in A^\#\setminus A$. Then $f(t) = b$ and $f(\varphi(t)) = c$ or $-c$.

ii) There exists a nonlimit ordinal α such that $t \in A_\alpha \setminus A_{\alpha-1}$. Then $\varphi(t) \in A_{\alpha-1}$ and so $f(\varphi(t)) = -f(t)$.

iii) There exists a limit ordinal α such that $t \in \mathrm{cl}(\bigcup\limits_{\beta<\alpha} A_\beta)\setminus \bigcup\limits_{\beta<\alpha} A_\beta$. In this case there exists a net $\{t_\gamma\} \subset \bigcup\limits_{\beta<\alpha} A_\beta \setminus A$ such that $t_\gamma \to t$. Then $\varphi(t_\gamma) \in A^\#$ for each γ and so $f(t_\gamma) = -f(\varphi(t_\gamma))$, implying that $f(t) = -f(\varphi(t))$. This finishes the proof for $n = 1$.

Now let us fix some $n > 1$ and assume that the conclusion of the lemma is true for each $k < n$.†

Let $A = \bigcup\limits_{j=1}^{n} A_j$, where A_j are pairwise disjoint clopen sets satisfying $\varphi(A_j) \cap A_j = \emptyset$ for each $j = 1, 2, \ldots, n$. We define new sets $D_1, \ldots, D_n$ in the following way.

$$D_1 = A_1^\#, D_2 = (A_2\setminus D_1)^\#, \ldots, D_n = \left(A_n \setminus \bigcup_{j=1}^{n-1} D_j\right)^\# .$$

Notice that each D_j is a clopen set. Let $m \leq n$ be the largest number such that $D_m \neq \emptyset$. Obviously $A^\# = \bigcup\limits_{j=1}^{m} D_j$. Finally, introduce $E = A^\#\setminus D_m$. We claim that $\varphi^{-1}(E) \subset E$. Indeed assume, contrary to what we claim, that there exists a point

† To be precise, our induction hypothesis says that a desired function $f \in C(E^\#)$ exists for each clopen subset E of K satisfying the following conditions: $E \cap F = \emptyset$ and $E = \cup_{j=1}^{k} E_j$, where $k < n$ and $\{E_j\}_{j=1}^{k}$ are pairwise disjoint clopen subsets such that $\varphi(E_j) \cap E_j = \emptyset$ $(j = 1, 2, \ldots, k)$.

$t \notin E$ such that $\varphi(t) \in E$. Certainly $t \in A^{\#}$ because $E \subset A^{\#}$ and $\varphi^{-1}(A^{\#}) \subset A^{\#}$, and therefore $t \in A^{\#} \backslash E = D_m$.

Observe now that since $t \in D_m$, and $\varphi(t) \notin D_m$, we can conclude by Lemma A.1(3) that $t \in A_m \backslash \bigcup_{j=1}^{m-1} D_j$. In particular, $t \notin \bigcup_{j=1}^{m-1} D_j$. On the other hand, the conditions $\varphi(t) \in E \subset \bigcup_{j=1}^{m-1} D_j$ and $\varphi^{-1}(D_j) \subset D_j$, $j = 1, 2, \ldots, m-1$, imply that $t \in \bigcup_{j=1}^{m-1} \varphi^{-1}(D_j) \subseteq \bigcup_{j=1}^{m-1} D_j$, a contradiction.

A straightforward verification shows that $E = \left(\bigcup_{j=1}^{m-1} E \cap A_j \right)^{\#}$. Now, applying our induction hypothesis to D_m and E, we can find $f' \in C(D_m)$ and $f'' \in C(E)$ such that $f'(D_m) \cup f''(E) \subset \{b, c, -c\}$ and $f'(t) \neq f'(\varphi(t))$ whenever $t, \varphi(t) \in D_m$, and $f''(t) \neq f''(\varphi(t))$ whenever $t, \varphi(t) \in E$.

Finally, let $f(t) = f'(t)$ if $t \in D_m$ and $f(t) = f''(t)$ if $t \in E$. The function f is well defined and continuous on $A^{\#}$ since E and D_m are disjoint. It remains to notice that by Lemma A.1(4) we have $\varphi(D_m) \cap E = \varphi(E) \cap D_m = \emptyset$, and consequently the properties of f', f'' imply that $f(t) \neq f(\varphi(t))$, whenever t and $\varphi(t) \in A^{\#}$. ■

The next lemma supplements the previous one and it will be needed later on.

A.2.1. Lemma. *If A is a φ-invariant clopen subset of K and $f \in C(A)$ satisfies the following two conditions*

$$f(t) \in \{b, \pm c\} \quad \text{for each} \quad t \in A,$$

$$f(\varphi(t)) \neq f(t) \quad \text{for each} \quad t \in A,$$

then there exists an extension $g \in C(A^{\#})$ of f satisfying the same conditions for each $t \in A^{\#}$.

Proof. We use the notations introduced earlier. Since $\varphi(A) \subseteq A$, we obviously have that $\varphi(A_\alpha) \subseteq A_\alpha$ for each ordinal α. Now let us fix an α and suppose that we have already constructed an extension of f to $\bigcup_{\beta<\alpha} A_\beta$ with the desired properties. If

α is a limit ordinal, then f can be extended to $A_\alpha = \bigcup_{\beta<\alpha} A_\beta$ by continuity. If α is a non-limit ordinal, then $A_\alpha \setminus A_{\alpha-1}$ is a union of three pairwise disjoint clopen sets D_b, D_c and D_{-c} (some of them may be empty) such that $f(\varphi(t)) = b$ for each $t \in D_b$, $f(\varphi(t)) = c$ for each $t \in D_c$ and $f(\varphi(t)) = -c$ for each $t \in D_{-c}$. Letting $f \equiv -c$ on D_b, $f \equiv b$ on D_c and $f \equiv c$ on D_{-c} we obtain a continuous extension of f to A_α with the desired properties. ■

Since $\varphi(t) = t$ for each $t \in F$, the existence of a function f described in Lemma A.2 implies immediately the following corollary.

A.3. Corollary. *If A is a closed subset of K and $A \cap F = \emptyset$, then $A^{\#} \cap F = \emptyset$.*

Corollary A.3 is strong enough to imply one of the two main results of Z. Frolik which were used in Sections 12 and 13. For the convenience of the reader we give below the exact references to an easily available book by R. Walker [Wal] containing Frolik's results rather than to the original papers by Frolik [Fro1,2].

A.4. Corollary (Frolik). (Theorem on the existence of the invariant base [Wal, Theorem 6.34].) *For each $t \in F$ and for each neighborhood V of t there exists a clopen set W such that $t \in W \subseteq V$ and $\varphi(W) \subseteq W$, that is, there is a base of φ-invariant clopen neighborhoods of t.*

Proof. Without loss of generality we may assume that V is clopen. If $\varphi(V) \subseteq V$ there is nothing to prove. So assume that V is not φ-invariant and consider a set $A := V \cap \varphi^{-1}(K \setminus V)$. Clearly A is a clopen non-empty set and $A \cap F = \emptyset$. Therefore, by Corollary A.3, we have $A^{\#} \cap F = \emptyset$. It is plain to see that $W = V \setminus A^{\#}$ is the desired φ-invariant clopen neighborhood of t. ■

If a point $t \in K$ is not a fixed point of the mapping φ, then it certainly cannot have a base of φ-invariant neighborhoods. However if the point is periodic, then, as our next result shows, its orbit does have such a base.

A.4.1. Corollary. *For each periodic point $t \in K$ of period $p \in \mathbf{N}$ and for each neighborhood V of the orbit set $O_t = \{t, \varphi(t), \cdots, \varphi^{p-1}(t)\}$ there exists a clopen set W*

such that $O_t \subset W \subseteq V$ and $\varphi(W) \subseteq W$, that is, there is a base of φ-invariant clopen neighborhoods of O_t.

Proof. Since t is a fixed point of the mapping φ^p, by Corollary A.4 we can find a clopen neighborhood W_0 of t such that $\varphi^p(W_0) \subseteq W_0$ and $D_j = \varphi^j(W_0) \subseteq V$ for each $j = 0, 1, \cdots, p-1$. Notice that $D_0 = W_0$, each D_j is closed and $\varphi^j(t) \in D_j$.

Let $W_{p-1} = \varphi^{-1}(W_0) \cap V$. It is a clopen neighborhood of D_{p-1} and $\varphi(W_{p-1}) \subseteq W_0$. Proceeding we will define consecutively $W_j = \varphi^{-1}(W_{j+1}) \cap V$ $(j = p-2, p-3, \cdots, 1)$, a clopen neighborhood of D_j satisfying $\varphi(W_j) \subseteq W_{j+1}$. Obviously the set $W = \bigcup_{j=0}^{p-1} W_j$ is the desired φ-invariant clopen neighborhood of O_t. ■

Notice that the previous corollary implies immediately that the same conclusion remains true for any point $t \in K$ with a finite orbit, that is, if $\varphi^n(t)$ is a periodic point for some $n \in \mathbf{N}$. However, for an arbitrary closed φ-invariant subset D of K it is not true in general that D has a base of φ-invariant clopen neighborhoods. It is an open problem to find non-trivial conditions ensuring the existence of such a base.

A.5. Corollary. *For each closed subset F_1 of the set F and for each neighborhood V of F_1 there exists a clopen set $W \subseteq K$ such that $F_1 \subseteq W \subseteq V$ and $\varphi(W) \subseteq W$.*

Proof. For each $t \in F_1$ there exists a clopen neighborhood $W(t)$ of t such that $W(t) \subseteq V$ and $\varphi(W(t)) \subseteq W(t)$. Letting $W = \bigcup_{j=1}^{k} W(t_j)$, where $\{W(t_j)\}_{j=1}^{k}$ is a finite subcover of F_1, we obtain the desired neighborhood. ■

It is clear that an analog of Corollary A.5 may be formulated for closed subsets of periodic points of a fixed period. To prove this, one simply uses Corollary A.4.1 instead of Corollary A.4.

Now we are in a position to prove the main result of this section.

A.6. Theorem. *The set $F^{\#}$, generated by the set F of all fixed points of φ, is clopen in K.*

Proof. Our first and principal step is to prove the next implication

$$\operatorname{int}(F^{\#}) = \emptyset \Longrightarrow F = \emptyset, \tag{1}$$

so assume that $\text{int}(F^{\#}) = \emptyset$. Consider a collection $\mathcal{G} = (G, g)$ of all pairs, where G is a clopen subset of K satisfying

$$G = \varphi^{-1}(G), \tag{2}$$

and g is a continuous function on G satisfying the following two conditions:

$$g(t) \in \{b, c, -c\}, \tag{3}$$

where $b, c \in \mathbf{C}, b \neq \pm c \neq 0$ are some fixed numbers, and

$$g(\varphi(t)) \neq g(t) \quad \text{for each} \quad t \in G. \tag{4}$$

Notice that (4) implies that $G \cap F = \emptyset$ for each $(G, g) \in \mathcal{G}$.

We can define a partial order on $\mathcal{G}$ by letting that $(G_1, g_1) < (G_2, g_2)$ if $G_1 \subset G_2$ and $g_2 \equiv g_1$ on G_1. We claim that there there exists a maximal element (H, h) in $\mathcal{G}$ with respect to the introduced order. To prove its existence it will suffice to verify that Zorn's lemma is applicable. Let $\{(G_\gamma, g_\gamma)\}$ be a chain in $\mathcal{G}$. On the open set $E = \cup_\gamma G_\gamma$ we can certainly define a continuous function e satisfying (3) and (4) by just letting $e(t) = g_\gamma(t)$ for $t \in G_\gamma$. It is also obvious that E satisfies (2). (However, E may not be clopen, and therefore we cannot assert that $(E, e) \in \mathcal{G}$.) Consider the closure $\overline{E}$ of the set E. It is a clopen set and there exists a continuous extension $\overline{e}$ of e from E to $\overline{E}$. Clearly $\overline{e}$ still satisfies (3) and (4) on $\overline{E}$. Again we cannot assert that $(\overline{E}, \overline{e}) \in \mathcal{G}$ since now the set $\varphi^{-1}(\overline{E})$ may be larger than $\overline{E}$.

However, since $\overline{E} \subseteq \varphi^{-1}(\overline{E})$, we see that φ maps $\overline{E}$ into itself, and therefore to the function $\overline{e} : \overline{E} \to \{b, c, -c\}$ we can apply Lemma A.2.1. By this lemma there exists a continuous extension g of $\overline{e}$ from $\overline{E}$ to $G = (\overline{E})^{\#}$ satisfying (3) and (4). The set G satisfies (2) by Lemma A.1(1), so indeed, (G, g) belongs to $\mathcal{G}$ and is an upper bound for $\{(G_\gamma, g_\gamma)\}$. This proves that the conditions of Zorn's lemma are fulfilled, and consequently there exists a maximal element (H, h) in $\mathcal{G}$.

We claim that $H = K$. This will prove (1), since, as noticed above, $H \cap F = \emptyset$. Assume contrary to what we claim that $H \neq K$. Take an arbitrary non-empty clopen subset A_0 of K such that $A_0 \cap (H \cup F^{\#}) = \emptyset$.

Clearly the closed set $\varphi(A_0)\backslash A_0$ does not intersect the set $H \cup F^{\#}$ (because $\varphi^{-1}(H) = H$ and $\varphi^{-1}(F^{\#}) = F^{\#}$), and therefore there exists a clopen set A_1 such that $\varphi(A_0)\backslash A_0 \subseteq A_1$ and $A_1 \cap (A_0 \cup H \cup F^{\#}) = \emptyset$. Now we can find a clopen set A_2 such that $\varphi(A_1)\backslash(A_0 \cup A_1) \subseteq A_2$ and $A_2 \cap (A_0 \cup A_1 \cup H \cup F^{\#}) = \emptyset$. Proceeding in the same way we obtain a sequence $A_0, A_1, A_2, \ldots$ of clopen subsets of K satisfying for each $n \in \mathbf{N}$ the following two properties:

$$A_n \cap (A_0 \cup \ldots \cup A_{n-1} \cup H \cup F^{\#}) = \emptyset$$

and

$$\varphi(\bigcup_{j=0}^{n-1} A_j) \subseteq \bigcup_{j=0}^{n} A_j.$$

Now let $B_0 = A_0^{\#}$, $B_1 = A_1^{\#}\backslash B_0, \ldots, B_n = A_n^{\#}\backslash(\bigcup_{j=0}^{n-1} B_j), \ldots$. Observe that the sets B_n have the following three properties.

$$B_n \cap B_j = \emptyset \quad \text{if} \quad n \neq j. \tag{5}$$

$$B_n \cap H = \emptyset, \; n = 0, 1, 2, \ldots \tag{6}$$

$$\varphi(B_n) \subset B_n \cup B_{n+1}, \; n = 0, 1, 2, \ldots \tag{7}$$

Property (5) is obvious. Since $A_n \cap H = \emptyset, B_n \subset A_n^{\#}$ and $\varphi^{-1}(H) = H$ we have (6). To prove (7), let us notice that for each $n \in \mathbf{N}$

$$\varphi(B_n) \subseteq \varphi(A_n^{\#}) \subseteq A_n^{\#} \cup \varphi(A_n) \subseteq A_n^{\#} \cup \bigcup_{j=1}^{n+1} A_j \subseteq \bigcup_{j=1}^{n+1} B_j. \tag{8}$$

On the other hand, if we assume that

$$t \in \varphi(B_n) \cap (\bigcup_{j=1}^{n-1} B_j) = \varphi(B_n) \cap (\bigcup_{j=1}^{n-1} A_j^{\#}),$$

then $t = \varphi(u)$ for some $u \in B_n$, and hence

$$u \in \varphi^{-1}(\bigcup_{j=1}^{n-1} A_j^{\#}) \subseteq (\bigcup_{j=1}^{n-1} A_j^{\#}) \subseteq \bigcup_{j=1}^{n-1} B_j,$$

contradicting (5). Therefore $\varphi(B_n) \cap (\bigcup_{j=1}^{n-1} B_j) = \emptyset$. In view of (8) it establishes (7).

By Lemma A.2, for each j there exists a function $f_j \in C(B_j)$ such that $f_j(B_j) \subseteq \{b_j, c_j, -c_j\} \subset \mathbf{C}$ and $f_j(t) \neq f_j(\varphi(t))$, whenever $t, \varphi(t) \in B_j$. Let $b_j = 2$ and $c_j = 6$ for each $j = 0, 1, \ldots$. Now we define a function f on $\bigcup_{j=0}^{\infty} B_j$ by letting

$$f(t) = \begin{cases} f_j(t) & \text{if } t \in B_j \text{ and } j = 0, 2, 4, \ldots \\ f_j(t)/2 & \text{if } t \in B_j \text{ and } j = 1, 3, 5, \ldots \end{cases}$$

As usual, f can be extended continuously to $B = \mathrm{cl}(\bigcup_{j=0}^{\infty} B_j)$, and obviously $\varphi(B) \subseteq B$ and $|f(t) - f(\varphi(t))| \geq 1$ for each $t \in B$. Therefore $B \cap F = \emptyset$, and Corollary A.3 implies that $B^{\#} \cap F = \emptyset$. It is plain to see that $\varphi(B^{\#}) \subseteq B^{\#}$ and also, in view of (6), that $B^{\#} \cap H = \emptyset$.

Applying again Lemma A.2 to the set $B^{\#}$, we can find a function $h' \in C(B^{\#})$ taking on values in $\{b, c, -c\}$ and satisfying $h'(\varphi(t)) \neq h'(t)$ for each $t \in B^{\#}$. Letting now $H_1 = H \cup B^{\#}$ and letting

$$h_1(t) = \begin{cases} h(t) & \text{if } t \in H \\ h'(t) & \text{if } t \in B^{\#} \end{cases}$$

we obtain a couple $(H_1, h_1) \in \mathcal{G}$ satisfying $(H_1, h_1) > (H, h)$, in spite of the maximality of (H, h). This concludes the proof of the first step.

Return now to the general case. We want to show that the set $F^{\#}$ is clopen. In view of Lemma A.1(6) it is enough to verify that $F \subseteq \mathrm{int}(F^{\#})$, that is, that the set $F_1 = F \backslash \mathrm{int}(F^{\#}) = \emptyset$. If $F_1 \neq 0$, then in view of Corollary A.5, there would exist a clopen set V_1 such that $F_1 \subseteq V_1$, $\varphi(V_1) \subseteq V_1$ and $V_1 \cap \mathrm{int}(F^{\#}) = \emptyset$. In particular we have

$$V_1 \cap \mathrm{int}(F_1^{\#}) = \emptyset. \tag{9}$$

Consider now $\varphi_1 = \varphi | V_1$. Clearly φ_1 is a continuous mapping of V_1 into V_1 having F_1 as the set of its fixed points. It is immediate from (9) that $\mathrm{int}(F_1^{\#_1}) = \emptyset$, where $\#_1$ denotes the #-operation on V_1 with respect to the mapping φ_1. Therefore, by (1), the set $F_1 = \emptyset$, a contradiction. ■

A.7. Corollary (Frolik). ([Wal, Theorem 6.25]). *If φ is a homeomorphism of K into itself, then the set F is clopen.*

Proof. Since φ is one-to-one, $\varphi^{-1}(F) \subseteq F$ and so by Lemma A.1(1) $F^{\#} = F$. By the previous theorem F is clopen. ■

It is an interesting open question whether we can replace in Frolik's theorem the assumption that φ is a homeomorphic embedding by the assumption that φ is open and continuous.† As the next result shows, the answer is affirmative under one additional assumption about K.

A.8. Corollary. *Let K be a hyperstonean compact space and φ an open continuous mapping of K into itself. Then the set F is clopen.*

Proof. First we will establish a fact which is of an independent interest and which asserts that whenever φ is open, there exists a simpler description of the operation $\#$. Namely, for each $A \subset K$ the identity $A^{\#} = \mathrm{cl}(\bigcup_{j=0}^{\infty} \varphi^{-j}(A))$ holds. Let us denote the latter set by A^{ω}. Clearly A^{ω} is closed and $A \subseteq A^{\omega} \subseteq A^{\#}$. Our claim will follow if we show that $\varphi^{-1}(A^{\omega}) \subseteq A^{\omega}$. Assume, contrary to what we claim, that there is a point $t \in \varphi^{-1}(A^{\omega}) \setminus A^{\omega}$. Take any clopen neighborhood V of t, such that $V \cap A^{\omega} = \emptyset$. Hence $\varphi(V)$ is a clopen neighborhood of $\varphi(t) \in A^{\omega}$, whence there exists a point $p \in V$ such that $\varphi(p) \in \bigcup_{j=0}^{\infty} \varphi^{-j}(A)$. However, the last set is clearly φ^{-1}-invariant, so the point $p \in A^{\omega}$, a contradiction.

Our second step is to prove the following implication:

$$F \neq \emptyset \implies \mathrm{int}(F) \neq \emptyset. \qquad (\star)$$

By Theorem A.6 and our assumption, $F^{\#}$ is a non-empty clopen set. Recall that K is hyperstonean, that is, each first category set in K is nowhere dense. Consequently the equality $F^{\#} = F^{\omega}$ established above implies that $\mathrm{int}(\varphi^{-j}(F)) \neq \emptyset$ for some $j \in \mathbf{N}$. But φ is open, and so $\mathrm{int}(F) \neq \emptyset$, as claimed.

† If we assume that φ is only continuous, then, as simple examples show, F may not be clopen.

The concluding part of the proof is similar to that of Theorem A.6. If F is not clopen, then $F_1 = F \setminus \mathrm{int}(F)$ is a non-empty nowhere dense subset of K, disjoint from the clopen set $\mathrm{int}(F)$. Therefore there is a clopen φ-invariant neighborhood V_1 of F_1, such that $V_1 \cap \mathrm{int}(F) = \emptyset$. Restricting φ to V_1 we arrive at a contradiction to $(\star)$. ■

Remark. In the course of the proof of the above corollary we have shown that $F^{\#} = F^{\omega}$ for each continuous open φ. Without the assumption that φ is open this result does not hold in general.

A.9. Corollary. (Lemma 12.6.2) *Let K be a hyperstonean compact space, E a clopen subset of K, and ψ an open continuous mapping of E into K. Then the set F of all fixed points of ψ is clopen.*

Proof. Consider $E_1 = \{t \in E : \psi(t) \in E\}$. Clearly E_1 is a clopen set and $F \subseteq E_1$. Introduce also a new space $K_1 = E_1 \oplus \{p\}$ which is a direct sum of E_1 and of a singleton $\{p\}$, and consider $\psi_1 : K_1 \to K_1$ defined as follows:

$$\psi_1(t) = \begin{cases} \psi(t) & \text{if } t \in E_1 \\ p & \text{if } t \in K_1 \setminus E_1. \end{cases}$$

It is plain to see that ψ_1 is an open continuous mapping of the hyperstonean space K_1 and the set F_1 of all fixed points of this mapping is $F \cup \{p\}$, which is clopen by the previous corollary. This implies clearly that F is also clopen. ■

Notice that if ψ is a homeomorphism, then the assumption that K is hyperstonean is redundant, since then we can use Corollary A.7 instead of Corollary A.8.

Recall [Wal, page 148] that a φ-invariant subset E of K is called **3-decomposable** if E is the pairwise disjoint union of three clopen subsets E_1, E_2, E_3 which satisfy $E_j \cap \varphi(E_j) = \emptyset$ for $j = 1, 2, 3$. In this case clearly, $E \cap F = \emptyset$.

The original proof by Frolik of the above Corollary A.7 depended on the existence of 3-decomposable subsets of $K \setminus F$. In Theorem A.10 which follows we will show that this result about 3-decomposability is in its turn a simple consequence of Corollary A.7; and so Corollary A.7 and Theorem A.10 are actually equivalent.

A.10. Theorem. *Let $\varphi : K \to K$ be a one-to-one continuous mapping of an extremally disconnected space K, and let F be the set of all fixed points of φ. Then the set $E = K \setminus F$ is 3-decomposable.*

Proof. The set E is clopen in view of Frolik's Theorem A.7, and E is φ-invariant since φ is one-to-one.

Let $\mathcal{G} = \{G\}$ be a collection of all those open subsets G of K for which $\varphi(G) \cap G = \emptyset$. Notice that for each $G \in \mathcal{G}$ the closure $\overline{G}$ also belongs to $\mathcal{G}$. Indeed, the set $\varphi(G)$ is open and disjoint from G, so their closures $\overline{G}$ and $\overline{\varphi(G)}$ are also disjoint since K is extremally disconnected. This implies that the clopen set $\overline{G}$ does belong to $\mathcal{G}$.

We order $\mathcal{G}$ by letting $G_1 > G_2$ iff $G_1 \supset G_2$. A straightforward verification shows that Zorn's lemma is applicable, and so there exists a maximal element $E_1 \in \mathcal{G}$ with respect to this order. In view of the previous part of the proof E_1 is a clopen set.

Now let $E_2 = \varphi(E_1)$ and $E_3 = E \setminus (E_1 \cup E_2)$. These sets are clopen and pairwise disjoint. It remains to notice that $E_j \cap \varphi(E_j) = \emptyset$ for $j = 1, 2, 3$. Indeed for $j = 1$ this follows directly from the definition, for $j = 2$ from our assumption that φ is one-to-one, and for $j = 3$ from the maximality of E_1. ■

We conclude with two more results about the properties of F.

A.11. Theorem. *If there exists a natural number n such that the cardinality of the inverse image of every point in K does not exceed n, that is,*

$$\operatorname{card}(\varphi^{-1}(t)) \leq n \quad \text{for each} \quad t \in K,$$

then F is a clopen subset of K.

Proof. The proof is similar to that of Theorem A.6, so we will omit some details. Let $\mathcal{G}$ be a collection of all pairs (G, g), where G is a clopen subset of K, $g \in C(G)$, $g(G) \subseteq \{0, 1, 2, \ldots, n\}$, and $g(t) \neq g(\varphi(t))$ if $t, \varphi(t) \in G$.

We introduce a partial order on $\mathcal{G}$ as it was done in the proof of Theorem A.6. Let (H, h) be a maximal element in $\mathcal{G}$, and suppose that there exists a point $t \in K \setminus (H \cup F)$. For each $j = 0, 1, \ldots, n$ we denote by H_j the subset of H, on which the function h takes on the value j. Since $t \notin H \cup F$ and $\varphi(t) \notin H$ there exists a clopen neighborhood V of

t such that $V \cap (H \cup F) = \emptyset$ and $\varphi(V) \cap H = \emptyset$. Moreover, using the condition of the theorem that $\text{card}(\varphi^{-1}(t)) \leq n$ we can additionally assume that $\varphi^{-1}(V) \cap H_j = \emptyset$ for some $j \in \{0, 1, \ldots, n\}$.

Extending h to $H \cup V$ by letting $h \equiv j$ on V, we come to a contradiction to the maximality of (H, h). This proves that $H = K \setminus F$, whence F is a clopen set. ■

If the set F of the fixed points of φ is clopen in K, then certainly F is extremally disconnected in the induced topology. It is rather surprising that the same conclusion remains true even if F is not clopen.

A.12. Theorem. *Let $\varphi : K \to K$ be a continuous mapping of an extremally disconnected compact space K. Then the set F of all fixed points of the mapping φ is extremally disconnected in the topology induced on F from K.*

Proof. First observe that for any clopen subset U of F, the sets $U^\#$ and $(F \backslash U)^\#$ are clopen in K and do not intersect. Indeed, in view of Corollary A.5 we can easily find two disjoint clopen subsets V_1 and V_2 of K such that $U \subset V_1$, $F \backslash U \subset V_2$ and $\varphi(V_j) \subseteq V_j$, $j = 1, 2$. By Lemma A.1(5) we have $V_1^\# \cap V_2^\# = \emptyset$, and consequently $U^\# \cap (F \backslash U)^\# = \emptyset$. This and Theorem A.6 imply that the sets $U^\#$ and $(F \backslash U)^\#$ are clopen in K.

To prove that F is extremally disconnected we must show that if U_1 and U_2 are disjoint open subsets of F, then their closures in F are also disjoint.

Consider the sets $H_1 = \text{cl}(\bigcup_{W_1 \subset U_1} W_1^\#)$ and $H_2 = \text{cl}(\bigcup_{W_2 \subset U_2} W_2^\#)$, where the jth union is taken over all clopen subsets W_j of U_j ($j = 1, 2$). By the preceding part of the proof the open sets $\bigcup W_1^\#$ and $\bigcup W_2^\#$ are disjoint, so their closures H_1 and H_2 in the extremally disconnected space K are also disjoint. Since clearly, $\text{cl}_F(U_j) \subseteq H_j \cap F$ ($j = 1, 2$), we arrive at the desired conclusion that $\text{cl}_F(U_1) \cap \text{cl}_F(U_2) = \emptyset$. ■

B. Proofs of Lemmas 12.6.1 and 12.6.3

Throughout this section K denotes an **extremally disconnected** compact Hausdorff space, E denotes a **clopen** subset of K and φ is a **continuous open** mapping from E into K.

Recall that a point $t \in E$ has period $m \in \mathbf{N}$ if $\varphi(t) \in E$, $\varphi^2(t) = \varphi(\varphi(t)) \in E, \ldots, \varphi^{m-1}(t) \in E$, $\varphi^m(t) = t$ and $\varphi^r(t) \neq t$ for $r < m$. For each $m \in \mathbf{N}$ we denote by $F^{(m)}$ the set of all points in E with period less than or equal to m. Accordingly, the set $F^{(1)}$ is nothing else but the set F of all fixed points.

Recall that $\Gamma = \{z \in \mathbf{C} : |z| = 1\}$. For each $n \in \mathbf{N}$, $n \geq 2$, we will denote by Γ_n the subgroup of Γ consisting of all n^{th} roots of unity. Obviously $|\gamma' - \gamma''| < \frac{2\pi}{n}$ for each two consecutive elements $\gamma', \gamma'' \in \Gamma_n$.

B.1. Theorem. (Lemma 12.6.1) *Assume that* $F^{(m)} = \emptyset$. *Then for an arbitrary* $\alpha \in \Gamma$ *there exists a function* $f \in C(K)$ *such that* $|f| \equiv 1$ *and* $|f(\varphi(t)) - \alpha f(t)| < \frac{2\pi}{m+1}$ *for all* $t \in E$.

Proof. The proof of this theorem will be divided into 3 steps.

Step 1. Here we assume that $E = K$ and that there exists a clopen set V satisfying the next two properties:

(1) The sets $V, \varphi^1(V), \ldots, \varphi^m(V)$ are mutually disjoint, and

(2) $K = V^\omega$, where $V^\omega = \text{cl}(\bigcup_{j=0}^{\infty} \varphi^{-j}(V))$.

Notice that in view of (1), the sets V, $\varphi^{-1}(V), \ldots, \varphi^{-m}(V)$ are also mutually disjoint.

For each $n \in \mathbf{N}$, $n \geq m+1$, we introduce a clopen set V_n by the following formula:

$$V_n = (V \cap \varphi^{-n}(V)) \setminus \bigcup_{j=1}^{n-1} \varphi^{-j}(V).$$

Clearly V_n is the set of all those points in V which leave V under the action of φ and return to V exactly on n^{th} iteration. (Of course the previous formula for V_n makes sense for each $n \geq 2$, but in view of (1) $V_1 = V_2 = \ldots = V_m = \emptyset$.) An additional property of the sets V_n which we will need a bit later is the next one.

(3) The open set $\bigcup_{k=m+1}^{\infty} V_k$ is dense in V.

Indeed, if $t \in V \setminus \mathrm{cl}(\bigcup_{k=m+1}^{\infty} V_k)$, then the points $\varphi^j(t)$, $j \geq 1$ never return to V, that is, $(V \setminus \mathrm{cl}(\bigcup_{k=m+1}^{\infty} V_k)) \cap \bigcup_{j=1}^{\infty} \varphi^{-j}(V) = \emptyset$. However, by (2), the set $\bigcup_{j=1}^{\infty} \varphi^{-j}(V)$ is dense in K, and so we can conclude that $V = \mathrm{cl}(\bigcup_{k=m+1}^{\infty} \varphi^{-k}(V))$.

Now introduce also the following clopen set:

$$W_n = V \cup \bigcup_{k=m+1}^{n} \bigcup_{j=1}^{k-1} \varphi^j(V_k),$$

where $n = m+1, m+2, \ldots$. Obviously the sets W_n are increasing.

Our next move is to show that a desired function f may be defined on W_n. For each $n = m+1, m+2, \ldots$ we denote by α_n an element in Γ_n, for which $|\alpha - \alpha_n| = \min\{|\alpha - \alpha'| : \alpha' \in \Gamma_n\}$. Clearly there are at most two choices for α_n (depending on α) and for either of them $|\alpha - \alpha_n| < \frac{2\pi}{n} \leq \frac{2\pi}{m+1}$.

First we define f on W_{m+1} by the following formula:

$$f(t) = \begin{cases} 1 & \text{if } t \in V \\ \alpha_{m+1}^j & \text{if } t \in \varphi^j(V_{m+1}), \quad j = 1, 2, \ldots, m. \end{cases}$$

The process of constructing f on W_n is inductive, so assume that a desired f has been already constructed on W_n and we want to extend it to W_{n+1}, that is, to a new piece of the trajectory generated by V_{n+1}.

Take any $x \in V_{n+1}$ and denote by $\ell(x)$ the smallest natural number such that $\varphi^{\ell(x)}(x) \in W_n$.

From (1), from the definition of W_n, and the fact that x returns to V exactly after $(n+1)$ iterations we can easily conclude that

$$m + 1 \leq \ell(x) \leq n + 1.$$

Since the point $\varphi^{\ell(x)}(x) \in W_n$, our function f is defined at this point, and so the number $z_x = f(\varphi^{\ell(x)}(x)) \in \Gamma$. Consider the equation $\gamma^{\ell(x)} = z_x$. It has $\ell(x)$ solutions, and we denote by γ the one which is closest to α. Since $\ell(x) \geq m+1$ we have $|\gamma - \alpha| < \frac{2\pi}{m+1}$.

Now for each $1 \leq j \leq \ell(x)$ we let $f(\varphi^j(x)) = \gamma^j$, thus defining f at each point of the trajectory generated by x, which lies outside of W_n. Certainly f is continuous and satisfies

$$|f(\varphi(t)) - \alpha f(t)| \leq \frac{2\pi}{m+1} \quad \text{if } t, \varphi(t) \in W_{n+1}.$$

Proceeding in the same way we define a continuous unimodular function f on the set

$$W_\infty = V \cup \bigcup_{k=m+1}^{\infty} \bigcup_{j=1}^{k-1} \varphi^j(V_k),$$

for which

$$|f(\varphi(t)) - \alpha f(t)| \leq \frac{2\pi}{m+1} \quad \text{if } t, \varphi(t) \in W_\infty.$$

Consider a clopen set $U := \mathrm{cl}(W_\infty)$. In view of (3) the following equality holds:

$$U = \mathrm{cl}(\bigcup_{k=m+1}^{\infty} \bigcup_{j=0}^{k-1} \varphi^j(V_k)),$$

whence obviously $\varphi(U) \subseteq U$. Extending f continuously from W_∞ to U, we have that

$$|f(\varphi(t)) - \alpha f(t)| \leq \frac{2\pi}{m+1}, \quad t \in U.$$

We can naturally extend f to $\bigcup_{j=0}^{\infty} \varphi^{-1}(U)$ by letting

$$f(t) = \alpha^{-j} f(\varphi^j(t)), \quad t \in \varphi^{-j}(U) \backslash \varphi^{-j+1}(U), \quad j = 1, 2, \ldots.$$

Finally, we can continuously extend f to $\mathrm{cl}(\bigcup_{j=0}^{\infty} \varphi^{-j}(U)) = K$, and we see that

$$|f(\varphi(t)) - \alpha f(t)| \leq \frac{2\pi}{m+1}, \quad t \in K.$$

Step 2. Here we assume only that $E = K$. This case can be reduced to the previous one by a method similar to the proof of Lemma A.2.

Because $F^{(m)} = \emptyset$ we can certainly find, using the compactness of K, mutually disjoint clopen sets $V_1, \ldots, V_n$ such that $K = \bigcup_{k=1}^{n} V_k$ and also $\varphi^j(V_k) \cap \varphi^l(V_k) = \emptyset$ for

$1 \le k \le n$, $0 \le j < l \le m$. Let $W_1 = V_1^\omega, W_2 = (V_2 \backslash W_1)^\omega, \ldots, W_n = (V_n \backslash \bigcup_{k=1}^{n-1} W_k)^\omega$. Let k be the largest number such that $W_k \neq \emptyset$. Then $\varphi(W_k) \subseteq W_k = W_k^\omega$ and using Step 1 we can apply induction on n exactly as in the proof of Lemma A.2.

Step 3. We will reduce the general case to the previous one. Consider the direct sum $K_1 = K \oplus \{1, 2, \cdots, m+1\}$. Define now a mapping $\varphi_1 : K_1 \to K_1$ by the following formula:

$$\varphi_1(t) = \begin{cases} \varphi(t) & \text{if } t \in E \\ 1 & \text{if } t \in K \setminus E \\ t+1 & \text{if } t \in \{1, 2, \cdots, m\} \\ 1 & \text{if } t = m+1 \end{cases}$$

It is obvious that φ_1 is an open continuous mapping of the extremally disconnected space K_1 into itself. Clearly the set $F_{\varphi_1}^{(m)} = \emptyset$, so by Step 2, applied to $\varphi_1 : K_1 \to K_1$, there exists a unimodular function $g \in C(K_1)$ such that $|g(\varphi_1(t)) - \alpha g(t)| \le \frac{2\pi}{m+1}$ for each $t \in K_1$. It remains to take $f = g|K$, the restriction of g to K. ■

Now we are ready to prove the last result that was left in Section 12 without a proof.

B.2. Theorem. (Lemma 12.6.3) *Let K be hyperstonean and $\alpha \in \Gamma$. Then for each $m \in \mathbf{N}$ there exists a function $f \in C(K)$ such that $|f| \equiv 1$ on $K \backslash \varphi^{-m}(F^{(m)})$ and $|f(\varphi(t)) - \alpha f(t)| \le \frac{2\pi+1}{m}$ for all $t \in E$.*

Proof. The set $F^{(m)}$ is clopen by Corollary A.8. Let $K_1 = K \oplus \{1, 2, \cdots, m+1\}$ be again the direct sum of K and $\{1, 2, \cdots, m+1\}$, and let $\varphi_1 : E \oplus \mathbf{N} \to K_1$ be defined as follows:

$$\varphi_1(t) = \begin{cases} \varphi(t) & \text{if } t \in E \setminus F^{(m)} \\ 1 & \text{if } t \in F^{(m)} \\ t+1 & \text{if } t \in \{1, 2, \cdots, m+1\} \\ 1 & \text{if } t = m+1 \,. \end{cases}$$

Clearly φ_1 is a continuous mapping and $F_{\varphi_1}^{(m)} = \emptyset$. By the previous theorem there exists a unimodular function $g \in C(K_1)$ such that

$$|g(\varphi_1(t)) - \alpha g(t)| \le \frac{2\pi}{m+1}, \quad t \in E_1.$$

Finally, let us define a continuous function f on K by the next formula:

$$f(t) = \begin{cases} g(t) & \text{if } t \in K \setminus \varphi^{-m}(F^{(m)}) \\ \frac{j}{m+1} g(t) & \text{if } t \in \varphi^{-j}(F^{(m)}) \setminus \varphi^{-j+1}(F^{(m)}),\ j = 1, 2, \ldots, m. \\ 0 & \text{if } t \in F^{(m)}. \end{cases}$$

Clearly f is unimodular on $K \setminus \varphi^{-m}(F^{(m)})$ and for each $t \in E$

$$|f(\varphi(t)) - \alpha f(t)| \leq \frac{2\pi}{m+1} + \frac{1}{m+1} \leq \frac{2\pi + 1}{m}.$$

This concludes the proof. ■

References

[Abr1] Y. A. Abramovich, Multiplicative representation of operators preserving disjointness, *Proc. Netherl. Acad. Wetensch. Ser. A* **86** (1983), 265–279.

[Abr2] Y. A. Abramovich, Two results about the multiplicative representation of Riesz homomorphisms, *Univ. Craiova Mat. Fiz-Chim.* **14** (1986), 17–21.

[Abr3] Y. A. Abramovich, Some theorems about normed lattices, *Vestnik Leningr. Univ. Mat. Meh. Astronom. No.* **13** (1971), 5–11.

[AAB] Y. A. Abramovich, C. D. Aliprantis and O. Burkinshaw, On the spectral radius of positive operators, *Math. Z.*, 1992, forthcoming.

[AAK] Y. A. Abramovich, E. L. Arenson, and A. K. Kitover, Operators in Banach C(K)-modules and their spectral properties, *Dokl. Akad. Nauk USSR* **301** (1988), 525–528.

[AVK1] Y. A. Abramovich, A. I. Veksler, and A. V. Koldunov, Operators preserving disjointness, *Dokl. Akad. Nauk USSR* **248** (1979), 1033–1036.

[AVK2] Y. A. Abramovich, A. I. Veksler, and A. V. Koldunov, Operators preserving disjointness, their continuity and multiplicative representation, *Linear operators and their applications*, Sbornik Nauchn. Trudov, 1981, Leningrad, 13–34.

[AB1] C. D. Aliprantis and O. Burkinshaw, *Positive Operators*, Pure and Applied Mathematics Series # 119, Academic Press, New York & London, 1985.

[AB2] C. D. Aliprantis and O. Burkinshaw, *Locally solid Riesz spaces*, Pure and Applied Mathematics Series # 76, Academic Press, New York & London, 1978.

[AL] A. B. Antonevich and A.V. Lebedev, Spectral properties of operators with shift, *Izvestia Akad. Nauk USSR* **47** (1983), 915–941.

[Art] W. Arendt, Spectral properties of Lamperti operators, *Indiana Univ. Math. J.* **32** (1983), 199–215.

[AH] W. Arendt and D. R. Hart, The spectrum of quasi-invertible disjointness preserving operators, *J. Funct. Anal.*, **68** (1986), 149–167.

[AS] N. Aronszajn and P. Szeptycki, On general integral transformations, *Math. Ann.*, **163** (1966), 127–154.

[Arv] W. Arveson, *An invitation to C^*-algebras,* Springer, New York-Heidelberg-Berlin, 1976.

[Coe1] C. C. Cowen, Composition operators on H^2, *J. Operator Theory* **9** (1983), 77–106.

[Coe2] C. C. Cowen, Linear fractional composition operators on H^2, *Integral Equations Operator Theory* **11** (1988), 151–160.

[Dre] L. Drewnowski, Topological rings of sets, continuous set functions, integration.III, *Bull. Polish. Acad. Sci. Math.* **20** (1972), 439–445.

[DP] P. G. Dodds and B. de Pagter, Orthomorphisms and Boolean algebras of projections, *Math. Z.* **187** (1984), 361–381.

[DPR] P. G. Dodds, B. de Pagter, W. J. Ricker, Reflexivity and order properties of scalar-type spectral operators in locally convex spaces, *Trans. Amer. Math. Soc.* **293** (1986), 355–380.

[Dug] R. G. Douglas, *Banach algebras techniques in operator theory*, Academic Press, New York, 1972.

[DS1] N. Dunford and J. T. Schwartz, *Linear Operators, Part I*, Wiley-Inter science, New York & London, 1958.

[DS2] N. Dunford and J. T. Schwartz, *Linear Operators, Part III*, Wiley-Inter science, New York & London, 1971.

[Eva] R. Evans, Embedding $C(K)$ in $B(X)$, *Math. Scand.* **48** (1981), 119-136.

[Fro1] Z. Frolik, Fixed points of maps of extremally disconnected spaces and complete Boolean algebras, *Bull. Acad. Polon. Sci. Ser. Math. Astronom. Phys.* **16** (1968), 269–275.

[Fro2] Z. Frolik, Maps of extremally disconnected spaces, theory of types, and applications, *General Topology and its Relations to Modern Analysis and Algebra*, III (Proc. Conf., Kanpur, 1968), 131–142, Prague, Academia 1971.

[Gil] T. A. Gillespie, Boolean algebras of projections and reflexive algebras of operators, Proc. London Math. Soc. (3) **37** (1978), 56–74.

[GdR] A. Goullet de Rugy, La structure idéale des M-espaces, *J. Math. pures et apll.* **51** (1972), 331-373.

[HO1] D. Hadwin and M. Orhon, Reflexivity and approximate reflexivity for Boolean algebras of projections, *J. Funct. Anal.* **87** (1989), no.2, 348–358.

[HO2] D. Hadwin and M. Orhon, A noncommutative theory of Bade functionals, *Glasgow J. of Math.* **33** (1991), no.1, 73–81.

[Har] D. R. Hart, Some properties of disjointness preserving operators, *Indig. Math.* **47** (1985), 183–197.

[HP] C. B. Huijsmans and B. de Pagter, Disjointness preserving and diffuse operators, *Comp. Math.* **79** (1991), 351–374.

[Kai] S. Kaijser, Some representations of Banach lattices, *Ark. Mat.*, **16** (1978), 179–193.

[Kam1] H. Kamowitz, The spectra of composition operators on H^p, *J. Functional Analysis* **18** (1975), 132–150.

[Kam2] H. Kamowitz, Compact weighted endomorphisms of $C(X)$, *Proc. Amer. Math. Soc.* **83** (1981), 517–521.

[KA] L. V. Kantorovich and G. P. Akilov, *Functional Analysis*, Moscow, 1977.

[KVP] L. V. Kantorovich, B. Z. Vulikh, and A. G. Pinsker, *Functional analysis in partially ordered spaces*, GITTL, Moscow, 1950.

[KK] Y. I. Karlovich and V. T. Kravchenko, Singular integral equations with non Carleman shift on an open contour, *Diff. Eq.* **17** (1981), 2212–2223.

[Kit1] A. K. Kitover, Disjoint operators in Banach lattices, *Dokl. Akad. Nauk USSR* **250** (1980), 800–803.

[Kit2] A. K. Kitover, Operators preserving disjointness and the mappings of extremally disconnected compact spaces, *Optimizaziya*, no.40 (1987), 138–149.

[Kit3] A. K. Kitover, The spectral properties of weighted homomorphisms in algebras of continuous functions and their applications, *Zap. Nauchn. Sem. Leningrad. Otdel. Mat. Inst. Steklov (LOMI),* **107** (1982), 89–103.

[KS] A. G. Kusraev and V. Z. Strihzevsky, Lattice normed spaces and majorized operators, *Investigations into geometry and mathematical analysis, Trudy Instit. of Mathematics, Novosibirsk* **7** (1987), 132–158.

[Kut] S. S. Kutateladze, Support sets of sublinear operators, *Dokl. Acad. Nauk USSR,* **230** (1976), 1029–1032.

[Lat] Y. D. Latushkin, On integro-functional operators with not one-to-one shift, *Izvestia Akad. Nauk USSR* **45** (1981), 1241–1257.

[Lev] V. L. Levin, Tensor products and functors in the category of Banach spaces defined by Banach lattices, *Trudy Moskov. Mat. Ob–va* **20** (1969), 43–82.

[LT] J. Lindenstrauss and L. Tzafriri, *Classical Banach Spaces II,* Springer, New York-Heidelberg-Berlin, 1979.

[Lip] Z. Lipecki, Extension of vector lattice homomorphisms, *Proc. Amer. Math. Soc.* **79** (1980), 247–248.

[Loz] G. Ya. Lozanovsky, On some Banach lattices, *Sibirsk. Mat. Zh.* **10** (1969), 584–599.

[LS] W. A. J. Luxemburg and A. R. Schep, An extension theorem for Riesz homomorphisms, *Proc. Netherl. Acad. Wetensch. Ser.* **A 82** (1979), 145–154.

[LZ1] W. A. J. Luxemburg and A. C. Zaanen, Notes on Banach Function Spaces III, *Proc. Netherl. Acad. Wetensch. Ser.* **A 66** (1963), 239–250.

[LZ2] W. A. J. Luxemburg and A. C. Zaanen, Notes on Banach Function Spaces XI, *Proc. Netherl. Acad. Wetensch. Ser.* **A 67** (1964), 507–517.

[MaC1] B. D. MacCluer, Compact composition operators on $H^p(B_N)$, *Michigan Math. J.* **32** (1985), 237–248.

[MaC2] B. D. MacCluer, Components in the space of composition operators, *Integral Equations Operator Theory* **12** (1989), 725–738.

[MCS] B. D. MacCluer and J. H. Shapiro, Angular derivatives and compact composition operators on the Hardy and Bergman spaces, *Canad. J. Math.* **38** (1986), 878–906.

[MW] P. T. N. McPolin and A. W. Wickstead, The order boundedness of band preserving operators on uniformly complete vector lattices, *Math. Proc. Cambr. Phil. Soc.* **97**(1985), 481–487.

[Mey] M. Meyer, Le stabilisateur d'un espace vectoriel réticulé, *C.R. Acad. Sc. Paris* **(A) 283** (1976), 249–250.

[Mon] B. Montador, Spectra of composition operators on C[0,1], *Lecture Notes in Math.* **399**, Springer, New York & London, 1974.

[Mys] A. D. Myshkis, Certain problems of the theory of differential equations with deviating argument, *Uspehi Mat. Nauk* **32** (1974), 173–202.

[Orh1] M. Orhon, Boolean Algebras of commuting projections, *Math. Z.* **183** (1983), 531–537.

[Orh2] M. Orhon, On vector lattice representation of Banach $C(K)$-modules, 1990, Preprint.

[Pet] K. Petersen, The spectrum and commutant of a certain weighted translation operator, *Math. Scand.* **37** (1975), 297–306.

[Ral] C. Rall, Über Boolische Algebren von Projection, *Math. Z.* **153** (1977), 199–217.

[RS] W. J. Ricker and H. H. Schaefer, The uniformly closed algebra generated by a complete Boolean algebra of projections, *Math. Z.* **201** (1989), 429–439.

[Rud] W. Rudin, *Functional analysis*, McGraw-Hill Book Company, 1973.

[SS] R. J. Sakel and C. R. Sell, A spectral theory for linear differential systems, *J. of Diff. Equat.* **27** (1978), 320–328.

[Sar] D. Sarason, Invariant subspaces and unstarred operator algebras, *Pacific J. Math.* **17** (1966), 511–517.

[Sc] H. H. Schaefer, Spectral measures in locally convex algebras, *Acta Math.* **107** (1962), 125–173.

[Sch] H. H. Schaefer, *Banach Lattices and Positive Operators,* Springer, New York-Heidelberg-Berlin, 1974.

[Sha] J. H. Shapiro, The essential norm of a composition operator, *Annals of Math.*(2) **125** (1987), 375–404.

[ST] J. H. Shapiro and P. D. Taylor, Compact, nuclear, and Hilbert-Schmidt composition operators on H^2, *Indiana Univ. Math. J.* **23** (1973), 471–496.

[Tak] M. Takesaki, *Theory of operator algebras I*, Springer, New York-Heidelberg-Berlin, 1979.

[Vek] A. I. Veksler, *Cyclic Banach spaces and Banach lattices, Soviet Math. Dokl.* **14** (1973), 1773–1779.

[Vul] B. Z. Vulikh, *Introduction to the theory of partially ordered spaces,* Walters-Noordoff Sci. Publication, Groningen, 1967.

[Wal] R. G. Walker, *The Stone-Čech compactification,* Springer, New York-Heidelberg-Berlin, 1974.

[W] B. Walsh, Structure of spectral measures on locally convex spaces, *Trans. Amer. Math. Soc.* **120** (1965), 295–326.

[Wic1] A. W. Wickstead, Representation and duality of multiplication operators on Archimedean Riesz spaces, *Comp. Math.* **35** (1977), 225–238.

[Wic2] A. W. Wickstead, Banach lattices with trivial center, *Proc. Royal Irish Acad.* **88** (1988), 71–83.

[Wic3] A. W. Wickstead, The ideal center of a Banach lattice, *Proc. Royal Irish Acad.* **76** (1976), 15–23.

[Wil] W. Wils, The ideal center of partially ordered vector spaces, *Acta Math.* **127** (1971), 41–77.

[Zaa] A. C. Zaanen, *Riesz spaces II,* North Holland, Amsterdam, 1983.

List of Symbols

Index